Pocket Guide to Cancer

Origins and Revelations

POCKET GUIDES TO
BIOMEDICAL SCIENCES

Series Editor
Dongyou Liu

A Guide to AIDS
Omar Bagasra and Donald Gene Pace

Tumors & Cancers: Brain – Central Nervous System
Dongyou Liu

Tumors & Cancers: Head – Neck – Heart – Lung – Gut
Dongyou Liu

A Guide to Bioethics
Emmanuel A. Kornyo

Tumors and Cancers: Skin – Soft Tissue – Bone – Urogenitals
Dongyou Liu

POCKET GUIDES TO
BIOMEDICAL SCIENCES

https://www.crcpress.com/Pocket-Guides-to-Biomedical-Sciences/book-series/CRCPOCGUITOB

The ***Pocket Guides to Biomedical Sciences*** series is designed to provide a concise, state-of-the-art, and authoritative coverage on topics that are of interest to undergraduate and graduate students of biomedical majors, health professionals with limited time to conduct their own searches, and the general public who are seeking for reliable, trustworthy information in biomedical fields.

Pocket Guide to Cancer

Origins and Revelations

By

Melford John

CRC Press is an imprint of the
Taylor & Francis Group, an **informa** business

CRC Press
Taylor & Francis Group
6000 Broken Sound Parkway NW, Suite 300
Boca Raton, FL 33487-2742

CRC Press is an imprint of Taylor & Francis Group, an Informa business

Printed on acid-free paper

International Standard Book Number-13: 978-1-138-74411-0 (Hardback)

Library of Congress Cataloging-in-Publication Data

Names: John, Melford, author.
Title: A guide to cancer : origins and revelations / Melford John.
Other titles: Pocket guides to biomedical sciences.
Description: Boca Raton : Taylor & Francis, 2018. | Series: Pocket guides to biomedical sciences
Identifiers: LCCN 2017021160| ISBN 9781138744134 (hardback : alk. paper) | ISBN 9781138744110 (pbk. : alk. paper)
Subjects: | MESH: Neoplasms | Handbooks
Classification: LCC RC263 | NLM QZ 39 | DDC 616.99/4--dc23
LC record available at https://lccn.loc.gov/2017021160

Visit the Taylor & Francis Web site at
http://www.taylorandfrancis.com

and the CRC Press Web site at
http://www.crcpress.com

I dedicate this book to my dear and loving wife Archna, who was brave or foolish enough to take me off the shelf, dust me off, and apply a bit of polish to take off some of the rough edges.

Contents

Series Preface

Dramatic breakthroughs and non-stop discoveries have rendered biomedicine increasingly relevant to everyday life. Keeping pace with all these advances is a daunting task, even for active researchers. There is an obvious demand for succinct reviews and synthetic summaries of biomedical topics for graduate students, undergraduates, faculty, biomedical researchers, medical professionals, science policy makers, and the general public.

Recognizing this pressing need, CRC Press has established the *Pocket Guides to Biomedical Science* series, with the main goal to provide state-of-the-art, authoritative reviews of far-ranging subjects in short, readable formats intended for a broad audience. Volumes in the series will address and integrate the principles and concepts of the natural sciences and liberal arts, especially those relating to biomedicine and human wellbeing. Future volumes will come from biochemistry, bioethics, cell biology, genetics, immunology, microbiology, molecular biology, neuroscience, oncology, parasitology, pathology, and virology, as well as other related disciplines.

Focusing on cancer, a disease that is constantly perplexing for some due to rapid advances being made in the field, and immensely terrifying for others with a mere mention given its destructive potential, the current volume presents a timely, state-of-the-art overview on key aspects of cancer biology, genesis, immunity, detection, biomarker identification, and therapeutic strategies. The goal of this volume is the same as the goal for the series—to simplify, summarize, and synthesize a complex topic so that readers can reach to the core of matter without the necessity to carry out their own time-consuming literature searches.

We welcome suggestions and recommendations from readers and members of the biomedical community for future topics in the series and experts as potential volume authors/editors.

Dongyou Liu, PhD
Sydney, Australia

Preface

Cancer presents mankind with our greatest medical challenge. It is enmeshed within a matrix of complexities, unlike any other disease or affliction. With the recent sequencing of the human genome and the ongoing unraveling of biological pathways, we are only now appreciating how truly complex and fascinating cancer really is. In writing this book, my aim is to immerse the reader in the intriguing world of war, corrupted pathways, forged relationships, betrayal, immortality, and cell death.

This book is written for the quintessential layperson standing at the doorstep of cancer genomics and bioinformatics taking a hesitant peek in, wondering if it is at all safe to enter. In it, I take an array of complex topics and break them down in clear and concise terms, so that anyone with a basic knowledge of science can understand. As with any of my lectures, it is my explicit aim, starting from first principles, to provoke, educate, and entertain. Anyone seeking light and understanding of the underlying causes of cancer and how different treatments work will find value and enlightenment in these pages. Unlike any other, this book provides an account of cancer from a fundamentalist biochemist's point of view.

Acknowledgment

I am thankful for and appreciative of my teachers and an old school British education system as it was when I was an impressionable student back in the day. How times have changed.

Author

Melford John is a senior lecturer at the University of the West Indies (UWI) in Trinidad and Tobago, where he has been teaching biochemistry to students of the Faculty of Medical Sciences for the past six years. He is engaged in research aimed at diagnosing cancer type using gene mutation and gene expression data. He also is working on the identification of cancer driver genes using protein interaction networks and gene expression data, the aim of which is to provide new targets for drug development. Prior to his current position at UWI, Dr. John was employed for nine years at the European Bioinformatics Institute (EBI) in Hinxton, United Kingdom, which is an arm of the European Molecular Biology Laboratory. At EBI, Dr. John worked for the Protein Data Bank in Europe group, engaged in the construction of databases and web services on the three-dimensional structure of proteins. Dr. John obtained a BSc and a PhD in biochemistry from the University of London, United Kingdom.

He is the chairman of the Caribbean Cancer Research Initiative, a not-for-profit organization dedicated to improving cancer patient outcomes and preventing more cancers in the Caribbean population through research-based interventions.

1 Origins

Cancer is driven by genetic aberrations that occur spontaneously on a daily basis. It is more complex than any other disease and is our greatest medical challenge. In recent times, cancer has emerged as one of the leading causes of death in developed countries. According to figures from the American Cancer Society, 1630 Americans would have died each day from cancer in 2016, amounting to 595,690 for the whole year. Cancer is the second most common cause of death in the United States, exceeded only by heart disease, and accounts for nearly one in every four deaths. During the average person's lifetime in the United States, there is a 42% chance of becoming a victim of cancer. As genetic aberrations accumulate with time, the risk of developing a tumor increases with age. Approximately 70% of estimated deaths for the four major cancers (lung and bronchus, colon and rectum, breast, and prostate) occur past the age of 65. What do we know of cancer? Why is it so difficult to cure? How can we reduce the risk of it developing?

We begin by examining the origin of cancer, what causes it. To do so, we need to become familiar with some basic biological terms related to the fascinating topic of molecular genetics. Let us take a trip back in time starting with the pivotal contributions of a Greek philosopher, an Italian poet, an English naturalist, and an Austrian monk. This could be the beginning of a funny joke, "they all walked into a bar," but it's not. Stay with me, lend me your ears. Let us boldly distill the essence of their works, and marvel at the creativity and elegance of it all.

1.1 Charles Darwin

In 1859, Charles Darwin, an English naturalist, published *On the Origin of Species by Means of Natural Selection*, in which he proposed a theory of evolution that occurs by a process of natural selection. In a changing world, life forms compete for limited resources such as food, water, and shelter from predators. Darwin worked on his theory for 20 years and concluded, animals and plants that are better able to compete are more likely to survive and reproduce. Over a period of time, species that adapt to their changing environment survive, while those that fail become extinct. In short, survival of the fittest. Figure 1.1 shows photos of Charles Darwin and Gregor Mendel.

Figure 1.1 Charles Darwin and Gregor Mendel.

It's obvious once said, but to step back and observe nature's thriving, resplendent creatures, to take note of subtle differences, and from that, shape a paradigm that fundamentally conflicted with established beliefs of the day took a creative, unhindered mind. Although he didn't know it at the time, Darwin was not the first to suggest that man was descended from animals, nor was he the first to introduce survival of the fittest as a concept.

1.2 Anaximander

More than 2500 years ago, the ancient Greek philosopher Anaximander (610–546 BC), credited by scholars as one of the greatest minds that ever lived, presented an explanation of the origin of man based on observations of nature. He suggested life originated from moisture that covered the earth before the sun evaporated it, and the first animals were a kind of fish, with a thorny skin. He also proposed that humans were not present at the earliest stages of the earth, but arose from fish later.

Even though his ideas were influenced by religious and mythical abstractions particular to his time, Anaximander was one of the first to attempt to explain the origin and evolution of the cosmos based on natural laws. He proposed the abstraction of "the boundless" as the origin of all things, and is credited as the originator of the theory of an open universe, which eventually replaced the notion of the closed universe of the celestial vault. To illustrate how ahead of his time Anaximander was, by comparison the ancient Greek philosopher Empedocles (490–430 BC) suggested the world was made up of the four elements: fire, water, earth, and air more than 50 years later.

1.3 Titus Lucretius Carus

The writings of the ancient Roman poet and philosopher Titus Lucretius Carus (99–55 BC) made more than a passing reference to survival of the fittest over 2000 years ago. Lucretius' poem, the title of which translates to

On the Nature of Things or *On the Coming into Being of Things*, consists of 7400 lines that extoll his philosophy on a wide range of topics including pleasure, religion, death, the physical world, and the nature of diseases.

He proposed a theory of evolution in which plants and animals were born out of the earth by the random combination of elements, after which the formation of new species ended. Natural selection led to "the ages after monsters died." Those organisms that "preserved alive" did so because of their greater fitness to survive, exemplified by their "cunning, valor, speed of foot or wing," and because of their usefulness to man. Lucretius did not believe in the descent of a new species from previously existing ones as Darwin did, and denied that land-dwelling animals evolved from marine animals as Anaximander proposed.

> *... And in the ages after monsters died,*
> *Perforce there perished many a stock, unable*
> *By propagation to forge a progeny.*
> *For whatsoever creatures thou beholdest*
> *Breathing the breath of life, the same have been*
> *Even from their earliest age preserved alive*
> *By cunning, or by valor, or at least*
> *By speed of foot or wing. And many a stock*
> *Remaineth yet, because of use to man . . .*
>
> *On the Nature of Things* by Lucretius
> Translated by William Ellery Leonard

Lucretius considered people to be made up of the same matter as everything that surrounded them, such as the stars and seas. He proposed that we should follow the pursuit of pleasure and avoidance of pain, as an alternative to living in fear of the gods. Though the gods exist, he suggested they neither made nor manipulated the world, and lived unconcerned with human affairs. Lucretius believed human beings should conquer their fears, come to terms with the fact that all things are transitory, and embrace beauty and the pleasures the world offers.

1.4 Gregor Mendel

Around the same time that Darwin was pondering the infinite, Gregor Mendel conducted meticulous crossbreeding experiments with plants, on which many of the rules of genes and their inheritance were established, now referred to as Mendelian genetics. Although not known at the time, Mendel's rules though developed from work on plants, apply to all living things that reproduce via sexual means. Mendel discovered that when purebred white-flower pea plants were crossbred with purebred purple-flower pea plants, the result was not a mixture of white and purple flower pea plants; all offspring were purple-flower pea plants. This led to the conclusion that there had to be two different traits for the color of the flowers, only one of which was expressed. All the first-generation pea plants were purple because the trait for this color was dominant, while the trait for white was recessive. Mendel also found that although not present in the first generation of plants, white flowers appeared in later generations.

Mendel's work led to the following conclusions:

- The inheritance of a trait is determined by units or factors that are passed on from parents to offspring unchanged.
- An offspring inherits a single trait from each parent.
- A trait may not show up in an offspring, but may still be passed on to future generations.

Today we can explain Darwinian evolution using Mendelian genetics.

1.5 About genes

The Danish botanist Wilhelm Johannsen coined the term gene used to describe Mendelian units of heredity in 1909, some 50 years after Darwin's publication on natural selection. He also made the distinction between the outward appearance of an organism, its phenotype, and its genetic traits, its genotype. Thus, the color of a flower is its phenotype, while the gene that codes for it is its genotype.

Genes are linear sequences of the molecules, guanine, cytosine, thymine, and adenosine that are chemically bonded together. They are commonly abbreviated to G, C, T, and A, respectively. Collectively they are referred to as bases. The bases are members of a type of chemical known as deoxyribose nucleic acid (DNA). It is commonly used to refer to genetic material representing any number of bases or genes. Who would have thought that the magic building blocks of the code for life would be four simple molecules? Another, perhaps more primitive genetic material, is ribose nucleic acid (RNA). This is similar in structure to DNA in that it is made up of four bases, but the bases are slightly different.

Genes vary greatly in length, from around 500 bases for the smallest human gene to over 2.2 million bases for the largest. Due to their vital importance for survival of the species, they are stored in a separate compartment in cells, away from potentially harmful chemicals in a structure known as the nucleus. Every cell in our body has a nucleus that contains a copy of the full complement of our DNA. Every cell that is, except for red blood cells and cornified cells in the skin, hair, and nails. Human red blood cells get rid of their nuclei on maturation to reduce their size for ease of circulation, and to increase their capacity to carry oxygen.

1.6 Decoding life

The full complement of genes in any living creature is referred to as its genome. In other words, the template used to create it. The sequencing of the genomes of many forms of life is currently taking place. The first animal genome to be sequenced was the nematode worm in 1998, which is composed of 100 million bases coding for 19,000 genes. In a mere three years, the human genome was found to consist of 3.2 billion bases. The first draft of the sequence was published in 2001. It was originally estimated that the human genome would consist of 100,000 genes, but that figure has been revised downwards on a number of occasions to the point where it now stands at 19,000. Surprisingly, we have around the same number of genes as the humble nematode worm.

From observations of peas, to genes, to decoding life itself, the word "awesome" was created for just these circumstances. In the words of the song "Rally Round the West Indies," by David Rudder, "little keys can open up mighty doors." Mendel, one imagines, might be moved to sing along with, "little peas can open up mighty doors," with perhaps a celebratory tipple of wine. Or maybe he would have no time for such frivolity, what with peas to cross-pollinate, prayers, and the like.

The templates of life for the 8.7 million different life forms that exist on our planet are all made up of DNA. Some viruses use RNA, but I'm controversially excluding them from the club of life as they are unable to replicate on their own. If there was a constitution for life forms, this would be a blatant violation of the first paragraph of the first article: "no self-respecting life form shall employ the genetic services of another species to reproduce." Guardians of the fate would suffer intolerable outrage. Is nothing sacred? Unthinkable! Whatever next?

1.7 Back to Mendel's work

The only way to get white-flower pea plants would be if genes inherited from both parents were for white flowers. Sexual reproduction therefore permits the transfer of two copies of the same gene to an offspring that may or may not be for the same phenotype. Who would argue about the euphoric benefits of sexual reproduction, conceived in the fertile mind of a celibate monk? The combination of genes from both parents is what gifts us intelligence, athletic ability, susceptibility to cancer, and other defining characteristics. Just as our parents raised us, it's also their fault that we are so messed up. Today we know genes code for proteins that serve a variety of functions in our bodies that include:

- Building blocks of various tissues
- Enzymes that catalyze chemical reactions
- Transmission of signals in our brains

Proteins are made up of 21 different amino acids. They range in size from 51 amino acids for insulin to about 30,000 for titin. The average protein is hundreds of amino acids in length. We don't have a good estimate of the number of proteins in the human body, but it is well over 100,000. The most common protein is collagen, a fibrous protein that is present in all tissues and organs that provides structure and strength. Proteins are involved in everything that goes on in our bodies, from muscle movement, to fighting diseases, to thought processes.

1.8 About chromosomes

A cell, the smallest unit of tissue, is surrounded by a membrane, which encloses substructures that carry out various functions. Of most importance is the nucleus, which also is surrounded by a membrane and contains 23 pairs of large molecules called chromosomes. We inherit one of each pair from our mother and the other from our father. Thus, we have two copies of each gene, one from each parent. Our 19,000 pairs of genes are distributed unevenly among the 23 pairs of chromosomes. Figure 1.2 shows a pair of chromosomes, one of which is inherited from mom and the other from dad. They are joined together at the centromere. Genes account for roughly 2% of the total number of bases, while non-coding regions account for the other 98%.

Basic arithmetic tells us that if there are 19,000 genes and over 100,000 proteins, a single gene on average codes for more than five proteins. This is made possible through a process known as gene splicing, in which gene products are cut up and reassembled to form templates to synthesize different proteins.

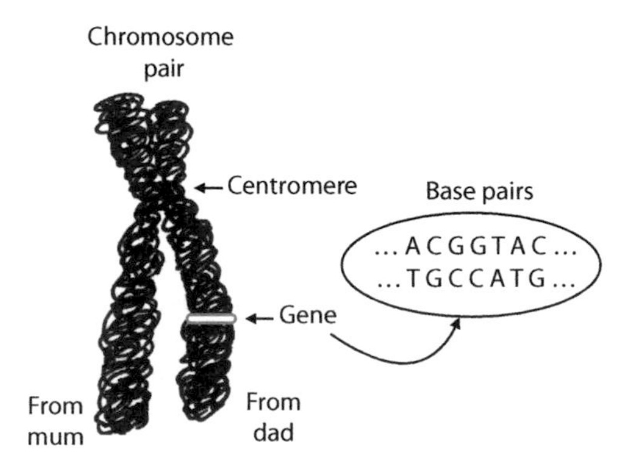

Figure 1.2 A pair of chromosomes. Each chromosome is made up of two strands of DNA, which is where the term "base pairs" comes from.

A chromosome is a very long molecule of DNA that is synthesized when cells divide by joining up millions of bases together. Each chromosome has a different number of genes in its sequence. For example, chromosome 1, the longest chromosome, has 249 million base pairs and over 3000 genes, while chromosome Y, the shortest, has closer to 200.

Chromosomes are numbered 1–22 from the largest to the smallest. In humans, the 23rd chromosome is made up of an XX pair in females, and an XY pair in males. The Y chromosome, which differentiates men from women, can only be inherited from the male parent. Cytoplasm, basically a gel consisting mainly of proteins and tiny organelles, is spread out in the space between the nucleus and the outer membrane.

It's natural to think that advanced species would have more genes and more chromosomes than primitive species, but not only is this incorrect, it is way off the mark. For example, the genome of the plant Paris japonica is 50 times longer than the human genome. The problem with so much baggage is that it takes a long time, and consumes a lot of resources, to duplicate a cell of Paris japonica during growth, which slows down the whole process. Plants with enormous genomes are more sensitive to pollution, and are at greater risk of extinction, because of the overhead involved in maintaining the integrity of their DNA. The tiny, adaptable water flea Daphnia pulex has 31,000 genes compared to 19,000 in humans. The reason for their adaptability relates to their high number of genes. It has been suggested that in response to changes in its environment, Daphnia pulex is able to switch on appropriate genes, altering its body to adapt.

1.9 Mutations

When a cell divides, its genome is duplicated so that it can be passed onto two daughter cells, a process known as mitosis. During this event, mistakes may be made resulting in a change in the base sequence of a gene. Such errors are termed mutations. Depending on the position at which they occur, a mutation may cause an impact on the function of a protein or make no difference. Mutations also arise spontaneously via attacks from chemicals, such as carcinogens in cigarette smoke.

1.10 Alleles

Mutations may give rise to alleles. Minor variations of the same gene that produce different phenotypes, such as purple flowers and white flowers, are termed alleles. We now know that a significant number of genes have multiple alleles, the frequencies of which vary from population to population. Alleles that are present in many healthy members of a species are considered normal. The threshold used to define normal is around 1%. Thus, a gene variant present in more than 1% of a population may be normal, whereas a gene variant present in less than 1% may be considered a mutated variant.

The genes that code for purple flowers and white flowers are alleles that are both normal. Alleles have different base sequences. New ones generally arise out of changes to genes that already exist. Natural selection results in discarding the majority of alleles produced by random mutations. In the full course of evolution, advantageous alleles increase in number and diversity, and the overall fitness of surviving species improves incrementally. Having had more time for them to develop, it follows that older species carry more distinct alleles in their genomes than younger ones. Chimpanzees, who have been around five million years longer than humans, have a far greater diversity in their alleles; yet we think they all look the same.

Mutations are caused by factors such as ultraviolet (UV) light, radiation, as well as external and internal chemicals that react with DNA. Every day, the DNA of our cells takes thousands of damaging hits. Why then do we not die out as the very fabric that codes for us is destroyed? Read on.

The integrity of DNA is important for the preservation of phenotype and the prevention of cancer, but maintaining it presents a huge challenge because:

- There is so much of it (3.2 billion base pairs).
- Each cell with a nucleus carries the full compliment.
- 200 billion new cells are produced every day.

1.11 DNA repair

To maintain DNA integrity, organisms have developed very sophisticated error-checking and repair mechanisms. The number of repair genes identified in humans currently stands at 176 or 0.9% of the genome. Many of these are common to different forms of life. In a primitive organism, such as yeast, 241 genes out of a total of 5600 are involved in the repair of DNA, which accounts for 4.3% of all genes. The importance of DNA repair has been recognized. The 2015 Nobel Prize in Chemistry was awarded to Tomas Lindahl from Sweden, Paul Modrich from the United States, and Aziz Sancar from Turkey for DNA repair discoveries. The rate of accumulation of mutations is accelerated if there are defects in any of the enzymes involved in the repair of DNA.

When a cell divides, during the initial pass of DNA duplication, one error occurs out of every 10^5 bases copied. A first proofreading step corrects this to reduce the error rate to one in 10^7. A second proofreading step reduces this to a more manageable one in 10^9. The net result of this is in the copying of the bases required for the production of a new cell, three errors are produced on average. Given the number of cells produced in an average lifetime (10^{16}), this means a total of 3×10^{16} mutations are produced.

1.12 What causes cancer?

With the passage of time, gene mutations and other alterations to DNA accumulate in all cells. The vast majority of these are lost when the cells they occur in die by natural means, or when they are repaired by repair enzymes. Cancer is caused by the accumulation of mutations in a cell that confer a growth advantage over surrounding cells. Only a small percentage of mutations drive the development of cancer. It is apparent that no single faulty gene or single genetic event causes cancer on its own. A full set of genetic defects required to cause cancer may take decades to accumulate by sporadic means. Therefore, the chances of getting cancer increase with age, particularly after 65.

1.13 Inheriting cancer

Although there is a perception that cancer is inherited, it must be made clear that this is not so. What is inherited is a genetic abnormality that predisposes its carrier to developing cancer. It's quite possible that someone born with such an abnormality may not ever become inflicted. A cautious lifestyle and a bit of luck helps. The consequences of inheriting a genetic disorder that drives cancer are:

- The risk of developing it is much higher than in the general population.
- The risk of it developing at an early age is higher.
- There is a greater risk of developing more than one cancer in the same tissue, or more than one type of cancer in the same person.

Genetic faults that assist the transformation of a normal cell into a tumor cell work in several different ways. They may:

- Signal cellular processes that initiate uncontrolled cell division
- Drive cellular processes that sustain uncontrolled cell division
- Accelerate the rate of accumulation of mutations

A faulty gene present at birth is potentially dangerous because it exists in virtually every cell, which at adulthood approximates to 10^{13} faulty genes in 10^{13} cells. These are either inherited from parents or occur spontaneously in a developing fetus. In contrast to congenital gene mutations, those that are formed after birth are present in only one cell. If the cell dies before becoming malignant, the mutation dies with it. These are relatively less dangerous. Depending on the function of a gene and the cell type it occurs in, a mutation may not be a problem, it may contribute to the development of cancer, or it may have serious deleterious consequences that even could be fatal. The silencing of the retinoblastoma 1 (RB1) gene is a case in point.

1.14 Retinoblastomas

The retina is nerve tissue that lines the back of each eye on which light falls and initiates vision. During the early stages of a baby's development, the number of retina cells increases very quickly, after which it stops increasing. A retinoblastoma is a rare cancer of the retina that is formed when both copies of the RB1 gene are faulty or missing. It affects approximately 300 children in the United States and 80 in the United Kingdom each year. Signs and symptoms include white pupils, eye pain, and redness. Retinoblastomas usually develop in early childhood, typically before the age of five. When diagnosed early and treated promptly, they are often curable. However, if left untreated they can spread to other parts of the body and become life threatening.

Retinoblastomas originate in two different ways. Congenital retinoblastoma, which comprises approximately 33% of cases, is formed when a child is born with a faulty or missing RB1 gene. Sporadic retinoblastoma, which comprises the other 67% of cases, is formed in cases where the RB1 gene becomes faulty or is lost after birth. About 90% of children born with a faulty RB1 gene go on to develop retinoblastoma by the age of five. In such children, there may be more than one tumor in an eye, and tumors may develop in both eyes. Survivors of congenital retinoblastoma are at greater risk of developing other cancers later in life.

A small percentage of retinoblastomas are caused by large deletions of chromosome 13 that remove the RB1 gene and other neighboring genes. Due to the loss of all these genes, children afflicted with this condition present with slow growth, intellectual disability, and abnormal facial features such as prominent eyebrows, a short nose, and ear abnormalities.

Congenital retinoblastomas differ from sporadic retinoblastomas in three different ways:

- They frequently develop in both eyes, unlike sporadic retinoblastomas, which typically develop in only one eye.
- Congenital retinoblastomas develop earlier than sporadic retinoblastomas.
- Children with congenital retinoblastoma have an increased risk of developing other forms of cancers, for example, a small number of children with congenital retinoblastoma also develop pineoblastoma, a tumor in the pineal gland at the base of the brain.

Why the differences between congenital and sporadic retinoblastomas? The RB1 gene codes for retinoblastoma protein (pRB), which serves a number of different functions that collectively suppress uncontrolled cell division and influence cell survival. We don't have a full understanding of how pRB works. The silencing of both copies of the RB1 gene in a cell by DNA modifications denies the cell the services of pRB and its suppressive effect on cell division. This unleashes uncontrolled cell division, which drives the development of retinoblastomas. Retina cells are particularly vulnerable to missing RB1 genes because they are heavily reliant on the services of pRB to suppress cell division, and because they divide quite rapidly early in life. Other cell types are not as reliant on pRB as retina cells, and so do not become cancerous at such an early age.

In cases of congenital retinoblastomas, one good gene and one bad gene are present in virtually every cell of the body, including millions of retina cells. The probability of one of the retina cells sustaining a hit that knocks out the only good RB1 gene, leaving two bad or missing copies, is very high, and is highest during early growth when millions of new cells are created. In a baby born with two good RB1 genes, it is highly unlikely both copies will get taken out in the same cell by two separate genetic events during the lifetime of the cell. For argument sake, if the probability of a single cell sustaining a single hit is one in a thousand, the chances of it sustaining two hits drops to one in a million. In the unlikely event of this happening, the chances of the same cell acquiring other mutations to fully transform into a cancer cell during its lifetime are extremely low. This is why sporadic retinoblastomas develop later than congenital retinoblastomas, and why they do not develop in both eyes.

1.15 Hereditary cancer

It is possible to develop cancer even if your parents never had it. A number of genetic, environmental, and lifestyle factors contribute to the risk of tumor formation. These factor in the development of:

- Hereditary cancers
- Familial cancers
- Sporadic cancers

Defective genes that are passed from parent to offspring are the main drivers of hereditary cancers. They tend to cluster in certain families in which they are more prevalent than in the broader population. Such cancer-causing genes are present in the sperm cells of males and the eggs of females. These germline mutations are dangerous because they occur in all cells of an offspring that has a nucleus, which is essentially every cell. Mutations that spontaneously occur in single cells after birth are termed somatic mutations and are not present in every cell.

Somatic mutations are not as dangerous as germline mutations because they arise out of a single event in a single cell. If the cell dies, the mutation dies with it. The mutation only lives if the cell avoids death, and only spreads if the cell multiplies. Thus, cancer-causing somatic mutations are far less dangerous than cancer-causing germline mutations. Having said that, somatic mutations that occur in the womb early in the development of an embryo may be just as dangerous as germline mutations. For example, it has been estimated that 75% of congenital retinoblastomas develop as a result of mutations in the womb sometime after conception.

Almost all germline mutations arise when cells fail to repair mistakes in their DNA. Older members of a population accumulate more mutations with time and therefore pass on more mutations to their offspring than younger ones. It has been shown that developing sperm and egg cells undergo the same mutational processes as other cells. It also has been estimated that fathers pass on nearly four times as many new mutations as mothers to their offspring. This is partly because women are born with all their eggs as immature cells, whereas sperm stem cells undergo division throughout their lives. Sperm cells also are produced in far greater numbers than egg cells, which means sperm stem cells go through more rounds of cell division than egg stem cells, accumulating more mutations in the process. By the age of 36, a man will pass on twice as many mutations to his offspring as he would at the age of 20, and by the age of 70, he would pass on eight times as many as he would at the age of 20. By starting families in their later years, parents run the risk of their children being born with genetic defects associated with diseases such as autism, schizophrenia, dwarfism, Down syndrome, and cancer. Men over the age of 50 are three times more likely to father a schizophrenic child than men under the age of 25.

Hereditary cancers account for a very small percentage of all cancers. About 5%–10% of breast and colon cancer cases can be linked to changes to single identifiable genes. Hereditary cancers develop relatively early in life, generally before the age of 50, often in children. The inheritance of defective genes shortens the time for the accumulation of the full complement of mutations required for the development of cancer. Families with defective genes tend to have multiple family members with the same or related cancers. Another common occurrence is two or more different cancers, such as colon cancer and breast cancer, in the same person. There also may be two or more cancers in the same organ, such as two separate colon cancers or two separate retinoblastomas.

The offspring of a parent with a faulty cancer-causing gene has a 50% chance of inheriting it. A child gets one of each member of a pair of chromosomes from its mother and the other from its father. Thus, if a faulty RB1 gene is on one of your father's two chromosomes, there is a 50% chance of you having it. An offspring that does not inherit a defective gene is free of risk.

1.16 Familial cancers

Familial cancers cluster in families, but do not necessarily follow patterns of inheritance. These are characterized by multiple cases of a specific type of cancer in a family that are higher than expected compared to the general population. Familial cancers are caused by the accumulation of mutation events induced by environmental and lifestyle factors. The overall risk of developing such cancers depends on the common genetic makeup of individuals within a family, and how they are influenced by long-term environment and lifestyle choices.

The risk of familial cancer is not passed on through families as a single defective gene, but by the accumulation of changes to many genes of somatic cells, each of which makes a contribution to its progression.

Factors associated with the accumulation of mutations and thus a higher risk of developing cancer include:

- Exposure to cigarette smoke
- Exposure to sunlight
- Exposure to radiation
- Excessive alcohol consumption
- Obesity
- A sedentary life style
- Excessive consumption of red meat
- Aging
- Infections of some viruses and bacteria

A study carried out by the European Prospective Investigation into Cancer and Nutrition analyzed the diets of over half a million Europeans, including 1300 who suffered from bowel cancer. One of the outcomes was the identification of an association of stomach and colon cancers with the consumption of large amounts of red meat.

1.17 Sporadic cancers

Sporadic cancers occur by sheer chance in individuals who have no known genetic risk factors or family history of cancer. Approximately 60% of cancers are of this type. Such cancers are not promoted by hereditary predispositions. Most of them occur late in life, and tend to occur in primary sites in which cancers are common, such as breast, prostate, lung, and colon. They also tend to occur in just the one primary site for any given individual. Where there are incidences of cancer in a family, there is no discernible pattern of inheritance.

The distinction between hereditary, familial, and sporadic cancer is important in the context of risk assessment, and is also of use in determining whether regular screening, risk-reducing surgery, or preventive chemotherapy are appropriate.

2

Why Cancer?

... It's a long, long road
From which there is no return
While we're on the way to there
Why not share ...

"He Ain't Heavy, He's My Brother" by The Hollies

The world we live in today is very different from when life first started. Between then and now many climatic eras have come and gone, accompanied by any number of extinct species. The essential drive to reproduce and the daily fight for survival, are encoded in the DNA of the cells of all forms of life. The internal processes whereby these are carried out started with early life and they have been improved upon ever since by natural selection, survival of the fittest. What has any of this to do with cancer? Read on.

2.1 Origins

The atmosphere of the Earth when it was formed, about 4.5 billion years ago, was very different to what it is now. It is believed that, like the atmospheres of Mars and Venus today, there was very little oxygen and a lot of carbon dioxide. It is likely that carbon dioxide arose from intense volcanic activity for the first billion years of the Earth's existence. As it cooled down, most of the water vapor in the air condensed and formed the oceans. We don't know how, but out of this torrid and unlikely setting life was born. The earliest recognizable life on Earth were single-cell organisms that appeared one billion years after the formation of the Earth, some 3.5 billion years ago, which are evident today in the form of fossils. They had the whole planet to themselves, in an environment that worked for them. Multicellular organisms appeared some three billion years later, around the same time oxygen became plentiful in the atmosphere. The significance of oxygen's contribution to life is apparent in every vital breath we take. It took three billion years to progress from the first single-cell life forms to the first multiple-cell life forms, and then only 0.5 billion years to get to where we are today.

The composition of the Earth's atmosphere currently consists of approximately 21% oxygen and 0.04% carbon dioxide. This incredible change is attributed to the consumption of carbon dioxide by plants and algae during photosynthesis, and the concomitant production of oxygen as a byproduct. Much of the early carbon dioxide ended up as fossil fuels in the form of oil and coal, which we are busy putting back into the air by burning the stuff.

Life on our planet has had the luxury of an enormous amount of time to develop, standardize, and fine tune biological processes to provide an impressive array of survival tools for the nine million different species living on it today. A significant number of the core processes that were in place very early on, such as the use of DNA as a template to code for life, cell division, photosynthesis, and glycolysis are still widespread among life forms, and remain pretty much unchanged. New features and innovations have been added by species after species, each trying to stay one step ahead of the competition and to keep abreast of an ever-changing environment. Once a species produces a gene that adds a new and useful feature, it becomes a permanent part of the gene pool for it and its descendants. Conversely, inferior genes that have been superseded are discarded or switched off. By the way, Homo sapiens have not added much by way of new genes to the pool, primarily because we haven't been around long enough.

All forms of life use the same four bases in their DNA, and pretty much the same basic set of 21 amino acids in their proteins, both of which suggest a common beginning for life on Earth. If other forms of life, using different molecules as building blocks were around in our history, they were unable to compete and are no longer with us. It is possible that life, if it exists elsewhere in the universe, may be based on other molecules. In which case, they too will need to undergo evolution and endure a survival of the fittest regime, the alternative being stagnation. They also may suffer from a form of cancer.

2.2 The pragmatic gene

A certain person has given genes a bad name by describing them as selfish. What does that make cancer? The persistence of genes that provide a survival advantage is eloquently demonstrated by the genetically inherited condition of sickle-cell anemia. Hemoglobin is a protein that carries oxygen around in the blood. It is made up of four protein molecules, two α-chains and two β-chains that cluster together to form a complex. Sufferers of sickle-cell anemia inherit faulty β-chain genes from both parents that impact the formation of the full hemoglobin complex. Instead of the four chains coming together to form a globular shape that is soluble, a long fibrous shape is formed that is not soluble. This happens because a gene mutation that brings about the substitution of a single amino acid on the β-chain of hemoglobin creates a sticky patch that causes it to aggregate, the effects of which are:

- Red blood cells adopt a sickle shape instead of a round doughnut shape.
- Sickle cells die within 10–20 days instead of 120 days like normal cells.
- Sickle cells clog up blood vessels causing pain, organ damage, and a lack of oxygen.

If untreated, sickle-cell anemia sufferers have a shorter life expectancy than normal. Additionally, in a competitive environment, the debilitating effect of the disease would be a distinct disadvantage in the context of survival of the fittest via natural selection. Yet, the gene persists to this day in some tropical countries. Why?

It turns out that carriers of a single faulty gene do not get sickle-cell anemia like those with two faulty genes. They are said to have the sickle-cell trait, which provides protection against malaria. They are 60% less likely to die from malaria than those who do not carry the sickle-cell trait. The frequency of the sickle-cell gene is more prevalent in areas where malaria is common. In such regions between 10%–40% of the population carry the sickle-cell trait. This figure seems about right because if everyone had two copies of the gene, everyone would have sickle-cell anemia and the population would struggle to survive. Alternatively, if very few had the faulty gene, very few would be protected from malaria, and would die of it. The case of sickle-cell anemia illustrates the preservation of genes that provide a survival advantage. It also shows a gene giving up its normal structure for the benefit of its carrier. If presented with the choice, a selfish gene would not choose to mutate, as it would no longer be itself. Genes are pragmatic.

2.3 Is evolution for real?

Despite overwhelming fossil evidence, and nascent genetic corroborations, Darwin's theory of evolution is controversial to this day. Undoubtedly, evolution conflicts with the belief that life was created in a matter of days. However, evolution and the existence of God are not mutually exclusive paradigms. Many a scientist has seen the photon. If God can create the heavens, earth, and women, evolution lies comfortably within His embrace.

There are those who preach religion on the one hand, and oppose the teaching of evolution on the other. The second irony about this is that religions are themselves subject to the forces of evolution. They compete for the same hearts and minds, without which there would be no donations to build places of worship and no jobs for celestial messengers, a stairway to extinction. The faith of the vast majority of religious followers is largely determined by an accident of birth. If you were born in a different family, the faith you adhere to so strongly could easily be quite different. We are prisoners of our own device.

... There's a lady who's sure all that glitters is gold
And she's buying the stairway to heaven
When she gets there she knows, if the stores are all closed
With a word she can get what she came for ...

"Stairway to Heaven" by Led Zeppelin

The difference between evolution and various religions is advocates of it have never killed anyone in its name or its defense. Scientists don't have thought police listening around corners, ready to brand you a heretic for speaking out against Newton's first law of motion or "the central dogma." God is perfectly capable of punishing those who trespass against Him, and may even seek to forgive those who trespass against Him as we seek to forgive those who trespass against us. Evolution is about adaptation, science is about understanding, while religions are about preservation and a stairway to heaven. The beauty of science and evolution is that they are based on observations of life and the universe supported by experimental evidence. The purity of science is that anyone can challenge and test it.

As a fundamentalist biochemist and man of the lab coat, versus a man of the cloth, science is my compass. Biochemistry, and what it stands for, point north. The chemistry of life, it doesn't get any more fundamental than that. However, my respect and admiration goes to physicists and their ever so tiny particles that keep getting smaller and smaller. It's only a matter of time before they discover absolutely nothing. I call this "The Melford Particle." The theory of evolution is strongly supported by science. The pieces scattered among various disciplines converge and fit. Geneticists, biologists, physicists, chemists, even anatomists, are essential branches of the same family. Let the truth and its nobility be your guide. It is the light at the end of the tunnel, not an oncoming train.

2.4 The driving force of cancer

The cardinal driver of evolution is the spontaneous mutations of genes. It is the same force that drives cancer. Instead of individual species competing for survival, individual cells compete for survival. Without mutations, there would be no changes to our genes, no acquiring of survival advantages, and therefore no evolution or cancer. At a higher level, two fundamental facets of survival of the fittest are the ability to reproduce at a faster rate than the competition, and an innate drive to spread out and find new habitats. This is apparent in primitive life forms, such as plants, and advanced forms, such as mammals. The prolific spreading of life over vast distances to discover new habitable environments selectively breeds the strongest. Even though many a seed will fall on many a stony ground, the strong will prevail and flourish. Cancer cells embrace the same philosophy when they pick up and migrate from their site of primary origin to distant regions of the body. Spread thy seeds.

Staying in one place is a good way to become extinct as food runs out and the sands shift beneath your feet. In a changing world, one must move to stand still. Man moved out of Africa 60,000 years ago, and look at us now. Landing on the moon was a small step for man, because we can't live there, a giant leap would be landing on a planet we can colonize. Plants, which are stationary by nature, have evolved the ability to disperse their seeds far and wide. Those that drop to the ground and germinate compete with their parents for minerals in the soil and sunlight. Fruits that carry the genetic code of plants in their seeds reward migratory animals with nourishment as they are consumed and the packaged seeds are distributed far and wide.

> *Your descendants will be like the dust of the earth,*
> *and you will spread out to the west and to the east, to the north and to the south.*
> *All peoples on earth will be blessed through you and your offspring.*
>
> Gen. 28:14

Overproduction and high attrition rates are part of the evolutionary process that helps to breed the fittest, and demonstrates that survival is not only about adaptation to a particular environment, but also about the search for new and fertile soil. A primitive species that can spread further and reproduce faster than a superior one can prevail by sheer weight of numbers.

2.5 Mutations change proteins

There is a fine balance between the accuracy of copying and repairing of damaged DNA and the demands of evolution. Out of necessity, life is dependent upon an optimum rate of mutation. If it is too slow and we stop evolving, this carries the risk of extinction in a changing environment. If it is too fast, our gene pool becomes unstable as good genes mutate before establishing themselves. In this regard, perfection is not a good thing. This is why we are not perfect. Stop blaming your parents, blame evolution. Species that get mutation and reproduction rates wrong are destined to vacate the planet.

When a gene mutates the sequence of its bases may be altered by:

- Insertion of one or more new bases
- Deletion of one or more bases
- Replacement of one or more bases

The linear sequences of the bases of genes are used as templates to synthesize proteins. DNA is first transcribed to form RNA, which is then translated to form proteins. This process was given the authoritative label of "the central dogma," an acute deviation from the stodgy practice of assigning unpronounceable Latin or Greek words to scientific discoveries. Three consecutive bases of DNA or RNA form a codon, each of which codes for a specific amino acid. As shown in Figure 2.1, during transcription the sequence of DNA codons of a gene is used as a template to synthesize a matching sequence of RNA codons. During translation, the sequence of RNA codons are translated one at a time into a sequence of amino acids to form proteins. A gene is said to be expressed when this happens. Single molecules of proteins are referred to as chains. These, as in the case of hemoglobin, sometimes aggregate to form protein complexes.

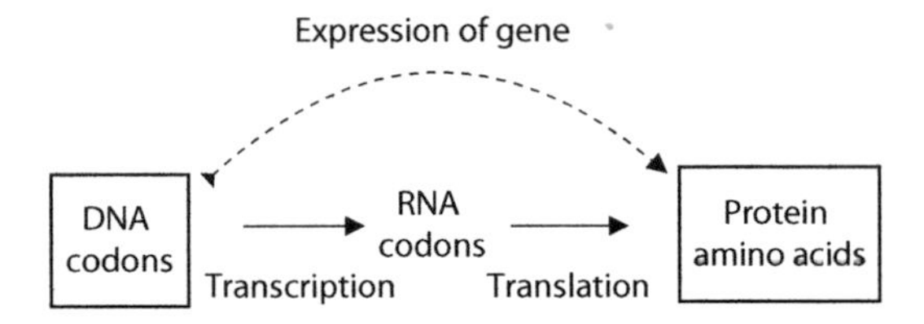

Figure 2.1 DNA is used as a template to synthesize proteins.

Although there are 21 different amino acids in proteins, the human genome only codes for 20. Hydroxyproline, which is abundant in collagen, is produced from proline. The choice of four possible bases for each of the three possible slots in a codon provides 64 possible outcomes (4 × 4 × 4). As there are only 20 amino acids to code for, each may be mapped by one or more codons. For example, six codons code for the amino acid arginine, while only one codes for tryptophan. There are three codons that code for a stop signal, which signals the endpoint of protein synthesis.

Mutations cause changes to codons, which cause changes in the amino acid sequence of proteins. The impact of a mutation on the protein being expressed varies from, no impact, to minor, to severe depending on the type of mutation and its location. The over expression or under expression of genes can bring about the same effect as activation or deactivation of proteins by mutations.

2.6 Architects of cancer

Each protein has a unique sequence of amino acids, which dictates its three-dimensional (3D) structure, a feature crucial to its function. A change in sequence is very likely to bring about a change in 3D structure, which in turn is very likely to prevent a protein from carrying out its true function. Depending on the role of the protein, mutations that impair its function may promote cancer, cause health problems, or even cause death. Therefore, it is of critical importance that proteins involved in functions such as cell division or repair of DNA are not mutated. The functions of proteins may be affected by inhibition of their activities or by rendering them in a permanently active or inactive state. If the altered function confers a growth advantage to a cell, the mutation is termed a driver mutation. Such proteins are the true architects of cancer.

Over the passage of time, which may be millions of years, the base components of genes may mutate, but sequences that code for amino acids, important for the function of proteins, remain the same. Mutations that drive the development of cancer tend to occur in such conserved regions, because they are most likely to impair the function of proteins.

2.7 Conservation of function

Examples of conserved regions of proteins and their relevance to their functions are abundant in nature. The conservation of the structure of insulin, throughout the course of evolution, provides a compelling case in point. Insulin is a very important hormone that, among other functions, helps to regulate the level of glucose circulation in the blood. It is first synthesized as inactive preproinsulin, a single molecule composed of 110 amino acids. Following synthesis, the first 24 amino acids of preproinsulin are cleaved off to form proinsulin (see Figure 2.2). Active insulin is formed from this by cutting off a piece in the middle at positions 54 and 89, amounting to the removal of

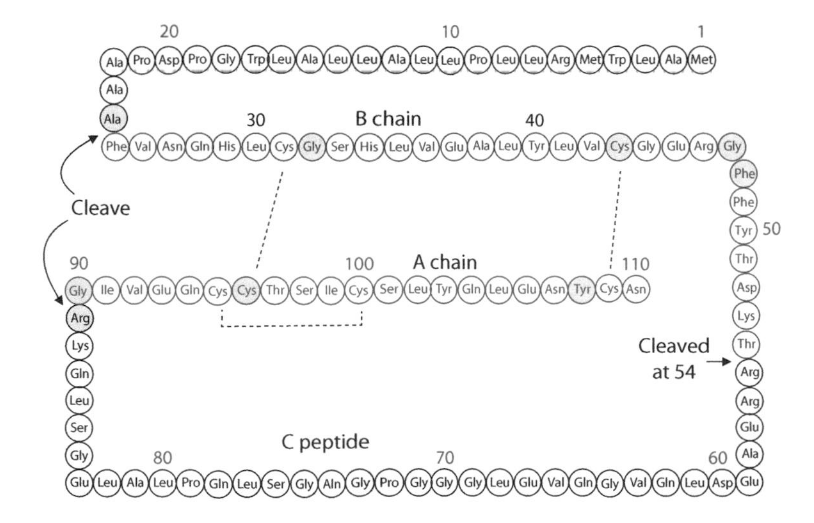

Figure 2.2 Preproinsulin is cleaved at amino acid 24 to form proinsulin, which is then cleaved at positions 54 and 89 to form insulin.

59 amino acids in total from preproinsulin (see Figure 2.2). This leaves active insulin composed of two chains, an A chain of 21 amino acids, and a B chain of 30 amino acids, joined together at two different points. The piece of preproinsulin that is cleaved off from the middle is referred to as the C peptide.

One would expect the 51 amino acids that form the A and B chains of insulin to be conserved to a greater degree than the discarded 35 amino acids of the C peptide. Structural comparisons of multiple species confirm this is true. For example, the A chain of pig insulin is exactly the same as the A chain of man, while the B chain of pig insulin differs from the B chain of human insulin by a single amino acid. Incredibly, the 51 amino acids of human insulin differ from pig insulin by a single amino acid, even though man diverged from pigs to form a separate species some 88 million years ago. In stark contrast, the 35 amino acids of the discarded C peptide of pig preproinsulin, is only 74% similar to the C peptide of humans. This demonstrates the region of preproinsulin that is required for activity, the insulin part, to be highly conserved, while other regions of the molecule not required for activity, such as the C peptide, are not highly conserved. The closer species are in evolutionary terms, the smaller the divergence in the structure of insulin and other important proteins.

Insulin was first used to treat diabetes in the 1920s, when it was commercially produced from the pancreas of slaughtered cows and pigs. However, the single amino acid difference in structure was enough to trigger allergic reactions in some patients. In 1978, the production of human insulin from yeast and bacteria through genetic engineering solved this problem. Today it also can be produced from plants. The process involves the isolation and incorporation of the insulin gene from humans into the DNA of the host being used as a factory.

There is very little variation in the structure of preproinsulin among primates, and none in the structure of insulin. The human preproinsulin molecule is the same as that of gorillas, even though we diverged to form a separate species eight million years ago. Also, the structure of preproinsulin in humans and other primates, including chimpanzees, gorillas, baboons, and monkeys, only varies by one or two amino acids. This level of conservation demonstrates the significance of the sequence of preproinsulin amino acids to the survival of primates. Unlike the preproinsulin molecule that shows some variation, the structure of insulin is the same for all primates studied thus far, and shows it to be highly conserved over the last 10–15 million years.

2.8 Conservation of functional components

There are nine sites of the preproinsulin molecule that when mutated are known to cause permanent neonatal diabetes. These are shaded in gray in

Figure 2.2. Of the nine sites, seven are part of the insulin molecule, while the other two are directly adjacent to it. One is adjacent to the beginning of the A chain, and the other is adjacent to the beginning of the B chain. In past times, individuals born with defective insulin molecules would not have survived long enough to pass them on, which accounts for their absence as alleles from our gene pool. This demonstrates that mutations of important functional components of proteins are more likely to be detrimental to life and illustrates why they are not conserved.

The pattern of an A chain and a B chain joined in the middle at two specific sites is not a regular feature of proteins in general, but is a characteristic feature of proteins of the insulin family. This signature motif has been conserved throughout evolution and is present in insulin molecules of primates, pigs, fishes, sea turtles, and insects, among others. The function served by insulin in humans today is quite different to the function it served when it first appeared. However, essential architectural components have persisted throughout evolution.

2.9 Protein families

During the course of evolution, the copying of existing genes to new locations in the genome, and their subsequent modification to serve new functions, or to serve the same function in different settings, is a method that appears to have been used to create new genes. This has given rise to the concept of protein families.

The protein families database, or Pfam, at the European Bioinformatics Institute, Hinxton, United Kingdom, is widely used and currently contains more than 16,000 protein families. Members of the same Pfam family share a common evolutionary history and therefore at least some sequential and functional aspects. Each family is characterized by functional units that are combined in different ways to generate proteins with unique functions. Proteins that share a common evolutionary origin and retain common structural elements are termed paralogues.

The widespread similarity of conserved domains among members of a protein family poses a problem for the pharmaceutical industry. Many faulty proteins contribute to the development of cancer. Suitable targets for cancer drugs are those involved in the transmission of signals between cells and within cells that regulate cell division and supportive metabolic pathways. Unfortunately, a significant number of these signaling proteins belong to a family of enzymes that share conserved regions, called tyrosine kinases. There are over 1000 kinases in humans, which makes it difficult to find a drug that targets a specific one. A drug that acts on a single protein is likely to cause fewer unpredictable side effects than one that acts on a large number of proteins.

2.10 Proteins that are enzymes

Enzymes are large protein molecules synthesized by all life forms for the specific purpose of making chemical reactions go faster that would otherwise not occur or take place too slowly. Life as we know it would not exist without them. Enzymes drive chemical reactions forward by lowering the energy required to kick-start them. Some work by binding to one or more reactants and altering their orientations or environment in a manner that makes the breaking and forming of chemical bonds easier. Sugar in a bowl has a lot of energy stored in it, which can be released by combustion, yet it does not burst into flames. This is because the activation energy required to start its combustion is high. A number of enzymes combust sugar in a controlled manner in our cells to provide energy. The pocket of an enzyme that reactants bind to is its active site. Typically, reactants fit active sites as a key fits into a lock. Active sites tend to be highly conserved. Mutations that cause any kind of change to them, or an area that provides access to them, are very likely to impact protein function.

The molecules that an enzyme binds to in the process of catalyzing a reaction is called its substrates. By the time a reaction is complete, the reacting substrates are converted to end products, and the enzyme molecule is reverted to its normal state, ready to catalyze a new reaction. Many drugs inhibit the activity of enzymes by blocking access of their natural substrates to their active sites. They can do this because they are similar in structure to the natural substrates. Such molecules are described as analogues. Most enzymes demonstrate a high degree of specificity toward their natural substrates. This is necessary to prevent them from catalyzing potentially harmful reactions.

2.11 Replacement mutations

There are many different types of mutations that can occur to cell DNA. The severity of the impact of mutations varies according to type. A replacement mutation is one in which a single base is replaced by another base in the DNA gene sequence. This contrasts with insertion and deletion mutations in which one or more bases are inserted into or deleted from the DNA gene sequence. The replacement of a base with another one creates a new codon. If it codes for a different amino acid, a single change in the amino acid sequence of the protein occurs. If this happens in a conserved region, and if the new amino acid is very different from the one it replaced, the effect on the function of the protein is likely to be deleterious. Replacement mutations may also prematurely terminate the synthesis of a protein if a stop codon is created by chance. More often than not, a truncated protein is bad news.

2.12 Frameshift mutations

Frameshift mutations arise when bases are inserted into or deleted from the DNA gene sequence, and the number of bases involved isn't a multiple of three. For instance, if a single base is deleted, the target codon and all subsequent codons become corrupted. Figure 2.3 shows how the deletion of G from the occurrence of the first ATG codon, which codes for the amino acid methionine, causes a change in sequence of the rest of the peptide. If the mutation occurs early in the protein sequence, a large proportion of incorrect amino acids will be incorporated in its structure, resulting in the synthesis of a new, random protein. Frameshift mutations produce abnormal proteins with unpredictable consequences that may be longer or shorter than the original protein. Thus, there are serious flaws in the intelligent design of our procreation that provide effective means of randomly corrupting proteins.

2.13 What drives cancer?

As previously stated, the driving force of cancer is the same as that of evolution, the need to mutate to survive. Cancer cells are normal cells that have become fitter by the accumulation of stochastic gene mutations, each conferring a survival advantage. They selfishly thrive at the cellular level, even though it is not in the best interest of their host. They cannot help it; it's in their DNA. Cancer is a side effect of the requirement of species to adapt, an affliction that surfs an evolutionary wave. The difference is that where evolution is spread across millions of years, the time scale for cancer is accelerated to decades.

Cancer is primarily a disease of the old. It takes time to acquire sufficient mutations to enable a transforming cell to overcome all the barriers in place to keep unregulated growth in check. Childhood cancers are biologically different from adult cancers, and tend to be simpler in terms of their genetic

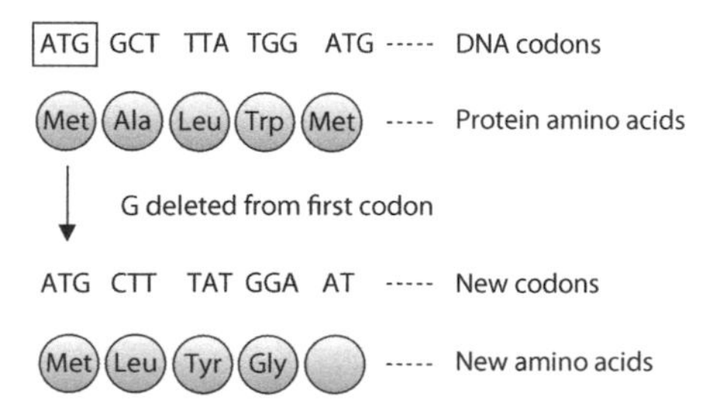

Figure 2.3 Deletion of G from first codon causes frameshift mutations with subsequent replacement of amino acids.

causes, which makes them more curable. Leukemias are the most common childhood cancers and are associated with abnormalities at the chromosomal level such as a missing copy, multiple copies, translocations, and missing portions.

There still is much to learn about the causes of different forms of cancer. For example, we don't know the complete profile of genetic mutations responsible for a particular type and how these change as the disease progresses. However, we do have a much better idea of the primary causes of some cancers.

How does the impairment of protein function cause cancer? To understand this, we need to take a look at cell division and how it gets out of control. Let us gaze upon the seminal contributions of Francis Crick, James Watson, and Rosalind Franklin.

2.14 Francis Crick, James Watson, and Rosalind Franklin

Based in Cambridge, England, some 44 years after the term gene was coined, Francis Crick, a British physicist, and James Watson, an American biologist, proposed a structure for DNA in a landmark paper published in *Nature* in 1953. In their work, the pair used the data of Rosalind Franklin, a British biophysicist and x-ray crystallographer. Her superb experimental work was a crucial contribution, yet she was given scant recognition at the time.

By 1962, when Crick, Watson, and Maurice Wilkins were awarded the Nobel Prize for physiology/medicine, sadly Franklin had died of cancer at the very young age of 37. The Nobel Prize only goes to living recipients, and can be shared among just three candidates. Franklin's contribution to the discovery of the structure of DNA is often overlooked. The title of Crick and Watson's landmark publication is *Molecular Structure of Nucleic Acids: A Structure for Deoxyribose Nucleic Acid*. Maurice Wilkins was a British-educated physicist who was born in New Zealand.

Crick and Watson proposed the structure of DNA to be a double helix, consisting of two long separate molecules attached to each other by weak hydrogen bonds. In the double helix, the base guanine on one strand always sits opposite the base cytosine on the other strand, while the base adenine always matches up with the base thymine, as shown in Figure 2.4. Each of the opposing pairs of bases is referred to as a base pair. Although the hydrogen bonds between each base pair are weak, the structure of DNA is stable because there are millions of them. This simple matching up is the key architectural component to the design of DNA.

The beauty of the double helix model of Crick and Watson is that it unlocked the mystery of how DNA is duplicated during cell division. It set the stage

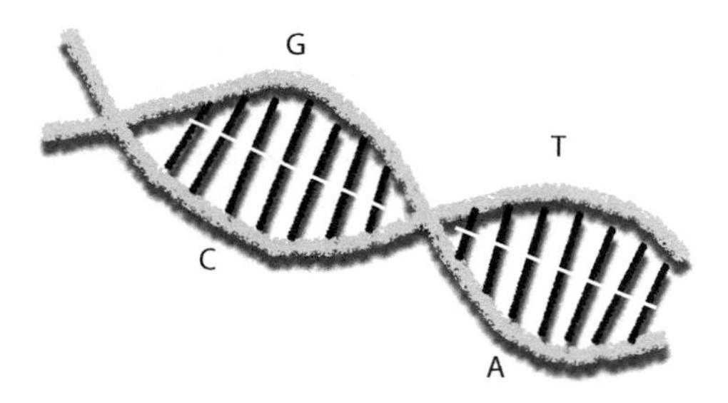

Figure 2.4 Double helix structure of DNA. Guanine sits opposite cytosine, and adenine sits opposite thymine.

for rapid advances in molecular genetics and laid the foundation for much of our understanding of how cancers are formed. As Sir Winston Churchill may have been moved to say, "Never was so much owed by so many to so few."

Because weak bonds attach base pairs on opposite strands of the helix, they are readily separated when they need to be for the purpose of duplication. In addition, because each base can only match up to its complement base, when the two strands of DNA are separated, each is able to act as a template for the synthesis of a new complementary strand. If ever there was an endorsement of the "keep it simple" school of thought in the design of systems, this is it.

There are 3.2 billion base pairs in our DNA, only 2% of which code for genes. Some of the other 98% code for features such as regulation of protein expression, but we don't fully understand the purpose of the rest, if indeed there are purposes to them. It is clear the areas that do not code for proteins, but are conserved over time, must serve a purpose.

The number of genes we have and the functions they serve is similar to that of other mammals, the mouse being a good example. Despite approximately 90 million years of independent evolution, the human and mouse genomes contain roughly the same number of protein-coding genes (19,000). Between mice and men, there is a gene overlap of approximately 75%.

The vast majority of sporadic mutations that occur over time are harmless, mainly because they occur in regions that do not code for genes. Of those that do, some may be in non-conserved regions, and therefore may not be important for the function of the protein they code for, and some may be in proteins that do not provide a growth advantage.

We can stop cancer if we can improve our DNA repair systems to the point where they fix mutations 100%. The problem with this is that it would stop us from evolving. Alternatively, we can improve our DNA repair system to the point that the time it takes for cancers to develop exceeds our natural lifespan. We may yet evolve to this state, but it won't be anytime soon.

There are probably many species alive and well today with DNA repair systems that delay the onset of cancer beyond their natural lifespans. There may even be some that have DNA repair systems that are 100% efficient. Scientists in the United Kingdom have recently estimated the Greenland shark to be the longest-living vertebrate on Earth. Radiocarbon dating of 28 of the animals suggested one female was about 400 years old. Their habitat is the cold, deep waters of the North Atlantic, in which they can be found swimming slowly. If you want to live long, slow down. The sharks grow up to five meters at a rate of one centimeter per year, and reach sexual maturity about the age of 150. Think of all those troubled teenage years. Why don't they start dying of cancer once past the age of 65 like us?

Our lifespan has only recently started to increase. It now averages above 80 years in some countries. Around the year 1800, life expectancy was fairly stable at around 40–45 years. If we survive other ravages of time, cancer will catch up with us.

> ... *Let's get together to fight this Holy Armageddon (One love)*
> *So when the Man comes there will be no, no doom (One song)*
> *Have pity on those whose chances grow thinner*
> *There ain't no hiding place from the Father of Creation...*
>
> "One Love" by Bob Marley

3
War on Cancer

Although, like other species, we regularly hunt and fight, war is an invention that is uniquely human. Whatever you may hear from politicians or historians, wars between nations are primarily fought to take control of land or other valuable resources belonging to someone else. "Give a dog a bad name and hang him" is a tactic that has been used repeatedly to provide a justifiable cause. Liberating a country equates to liberating its resources. The basic process is you have it, we want it, we take it. If you can't defend what is yours, you lose. It comes down to survival of the fittest. Governments often gain political capital from initiating wars, even if they make no sense in terms of national interests. You could say, money and power are the root of all wars. The citizens of a country almost always rally around its leader at a time of war, regardless of how foolish or unjust the cause, or how inept the leader. Many leaders experience a surge in popularity on a wave of nationalism and sense of collective duty during times of war against a common enemy led by an "evil" villain. It comes as no surprise that wars have been declared against all sorts of concepts, including cancer. Let's take a look at man and our wars from 36,000 feet. There may be some turbulence, so please don't leave your seat.

3.1 Fighting over differences

Genetically we are 99.9% alike, yet we discriminate and fight along lines of race, religion, and nationality. Of course, there is only one race, the human race, but there are those who wouldn't want you to think so. Our genes tell us the only 100% pure, modern-day humans are native Africans. If you were born outside of Africa, you are almost certainly up to 4% Neanderthal, a species unable to compete with *Homo sapiens*, and careless enough to find itself extinct. What would any aspiring modern-day devotees of the master race make of this? Hitler was right about there being a pure race. Ironically, he wasn't part of it. His words carried the weight of iron, but much of it was oxidized. Should the blood that flows through supreme veins be tainted with the blood of a species that swam in the shallow end of the gene pool? To find oneself up to 4% Neanderthal is careless, to be more than 95% African is wanton recklessness beyond the reaches of redemption.

We have been strutting our stuff around planet Earth for near on 200,000 years now, grabbing land and surviving communicable and non-communicable diseases as we relentlessly move on. Compared to other species that have been in existence for millions of years, we are new kids on the block. Dinosaurs appeared around 230 million years ago, and became extinct about 65 million years ago. They stalked the earth for 165 million years. Had it not been for their extinction, generally accepted as due to the fallout from a giant meteor, our ancestral linage may not have survived an onslaught from them and their kind. Since our divergence from chimpanzees, our nearest living relatives, some six million years ago, a number of hominids have come and gone. We are all that is left from our little branch of the tree of life. Nature offers no guarantee of survival, not even an extended warranty.

3.2 Out of Africa

Not content with our continent of birth, we boldly emerged out of Africa some 60,000 years ago to frequent other continents where no man went before. Figure 3.1 shows the path we took. On our way, we encountered and outlived Homo erectus. No mean feat, since our distant relative had been around for close to two million years. We also came across and outlived a more advanced coinhabitant in the form of Homo neanderthalensis, who, by comparison, had been around for a mere 300,000 years. For some unknown reason, Neanderthals became extinct about 28,000 years ago, even though they were stronger than us, and had bigger brains. Sometimes brains and brawn are just not enough. Nevertheless, Neanderthals were attractive enough for our ancestors to interbreed with them. Every form of refuge has its price.

In a short period of time, our intellect and dashing good looks have enabled us to live long and prosper, some benefiting more than others. The richest 1% of the world's population now owns 50% of global wealth. At one end of the scale are individuals with so much wealth they couldn't spend it all in a hundred lifetimes, while at the other end, there are children who go blind due to lack of proper nutrition. The wealthy exert and maintain a disproportionate effect on the way we, as a species, manage our affairs and our planet. They decide when we go to war and whom we fight against.

The jury is still out on our long-term survival of course, because we've only just begun. We are currently enjoying a fossil fuel party, melting the ice cap, using up resources, and spreading pollution like there is no tomorrow. There are now seven billion of us on Earth competing for resources that, as sure as day follows night, will run out. What happens when the music stops? Professor Stephen Hawking, a British theoretical physicist, who predicted that black holes emit radiation, suggested mankind has 1000 years to find a new planet to live on. He argues that, although the chance of a catastrophic disaster happening in any given year may be quite low, over a period of 1,000–10,000 years it becomes a near certainty. Nuclear wars,

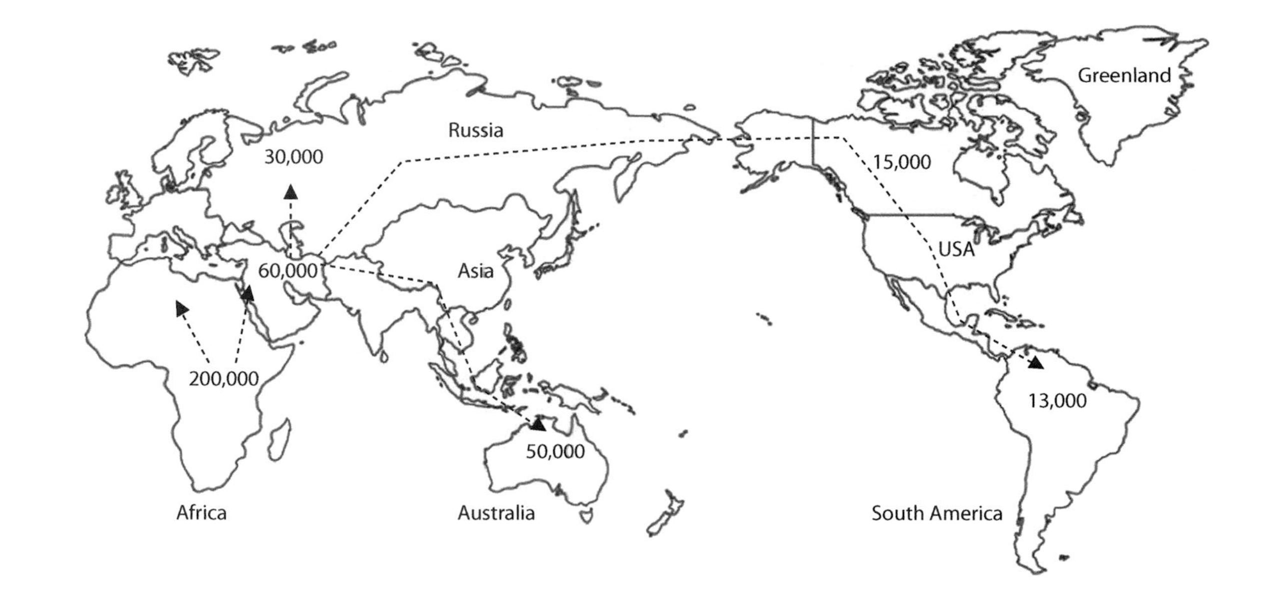

Figure 3.1 Migration of man out of Africa 60,000 years ago.

climate change, robots, and asteroids are all capable of wiping us out. Galaxies are moving further away from each other, and the speed at which they are moving away is increasing, so we may want to work together on an exit strategy at some point. As we individually strive to improve our personal lot in life, collectively we lose focus on the philosophic overview of what we are about and where we want to be. Short-term gains take precedence over long-term objectives.

3.3 Mortality

For now, due to advances in medical care, sanitation, and nutrition, particularly over the past 50 years, global life expectancy is on the increase. This is more marked in high-income countries that enjoy better all-round standards of living, organization, and education. A girl born today is expected to live some six years longer than one born a mere 30 years ago. This increase is caused primarily by reductions in deaths from communicable diseases and infant mortality. As a consequence, deaths from non-communicable diseases are on the rise. This trend is more pronounced in high-income countries where the top causes of deaths are now heart disease, cancer, and diabetes.

Mortality rates from communicable diseases caused by bacteria and viruses such as smallpox, the Plague, and tuberculosis that were once the scourge of mankind have declined dramatically. The most catastrophic pandemic in recorded history, the Spanish flu, which occurred in 1918, claimed the lives of over 50 million people. What stopped the influenza A virus from wiping us out completely? Due to genetic diversity among any large population of a species, there will be those who by chance have greater resistance to a particular pathogen compared to others. They live so that we, as a species, live to fight another day. In this context, the upkeep of genetic purity touted by supremists paves a clear path to extinction. The same strain of the influenza A virus, known as H1N1, claimed 18,000 victims when it returned in 2009 as swine flu. The annual number of people worldwide dying from acquired immune deficiency syndrome (AIDS) related causes steadily decreased from a peak of 2.3 million in 2005 to an estimated 1.6 million in 2012.

Like some cancers that develop resistance to drug treatment and relapse, some strains of pathogens mutate and become resistant to antibiotics, a consequence of evolution via natural selection. This phenomenon is an emerging public health issue. If not addressed, it could be catastrophic in the long run. We have won many a battle, but we may yet lose the war. Therefore, we need to be careful about the overuse of antibiotics and must discover new drugs with different modes of action. Studies have suggested that 30%–50% of antibiotics prescribed in hospitals are unnecessary or incorrect. Methicillin-resistant *Staphylococcus aureus* (MRSA) and clostridium difficile are frequently identified pathogens in U.S. hospitals and nursing homes. Each year, about 90,000 Americans become infected with MRSA, out of which 20,000 die, many of whom are children. *Clostridium difficile* has a higher infection rate, claiming 500,000 American sufferers and 29,000 deaths within 30 days of diagnosis.

3.4 Vaccinations

Although as recently as 1967, an estimated two million people died from smallpox, the disease was declared eradicated by the World Health Organization (WHO) in 1979, primarily because of the widespread use of vaccinations. Vaccines are prepared from pathogens, such as viruses and bacteria, by isolating fragments of them, or by preparing weakened versions of them. These do not cause an infection when introduced into a host, but elicit an immune response that protects against future infections. Our attempts at producing vaccines for cancer so far have been largely unsuccessful, but we do have one to counter metastatic prostate cancer. A major hurdle is the identification of biological material in cancer cells that differentiates them from normal cells. The cells of bacteria and viruses are very different to our cells, and so do not present this problem.

The WHO reports licensed vaccines are currently available to prevent, or contribute to the prevention and control of 25 different infections. This list includes tuberculosis, meningitis, poliomyelitis, diphtheria, hepatitis B, measles, mumps, pertussis, rubella, tetanus, and yellow fever. Having demonstrated our prowess in competing against communicable diseases and Neanderthals, how are we doing against non-communicable diseases? In particular, how are we doing against cancer?

3.5 Cancer

The news on cancer is not good. Globally it is a very common disease and incidence continues to rise. Most treatment regimens employed against cancer fail to cure it. This is in stark contrast to the widespread use of antibiotics and vaccinations that have been successfully curing a host of infections for many years. Once eliminated, these pathogens do not come back the way cancer does.

According to the American Cancer Society, the total cancer death rate rose for most of the last century, reaching a peak in 1991. The fall in death rates and improved survival rates are primarily attributable to reductions in smoking and improvements in early detection and treatment. Death rates are declining for the four most common types of cancer: lung, colorectal, breast, and prostate.

The WHO reports that non-communicable diseases were responsible for 68% of all deaths globally in 2012, up from 60% in 2000. The four main non-communicable diseases are cardiovascular disease, cancer, diabetes, and chronic lung disease. According to the American Cancer Society 42% of men and 38% of women have a lifetime risk of developing cancer. The figures are worse for the United Kingdom, where Cancer Research U.K. predicts that 54% of men and 48% of women will get cancer at some point in their lives. Cancer is on the brink of surpassing cardiovascular disease to become the leading killer in developed countries. The mere fact that we are living longer, and surviving cancer longer, means it is now present in more of us. It is destined in one form or another to touch the lives of all of us. Survival rates against the different forms of cancer are increasing due to:

- Improvements in medical treatment
- Better nutrition
- Changes in lifestyle

Growing tumors kill by interfering with one or more essential body functions. In lung tissue, they block absorption of oxygen and release of carbon dioxide, which if severe enough will cause death. Cancers may deplete the capability of the immune system so that infections become lethal.

Those unfortunate enough to be afflicted with cancer endure years of suffering at great emotional and financial cost. It is in the interests of the people of nations to reduce their cancer burden. At the end of 1971, U.S. President Richard Nixon signed the National Cancer Act. An act generally viewed as the beginning of the "war on cancer" (though it was not described as such in the legislation). However, in his state of the union address earlier that year President Nixon referred to the "national commitment for the conquest of cancer."

Undoubtedly, placing cancer in the public eye intensified research efforts, but the use of the emotive term "war" probably focused attention on winning the fight, rather than on prevention. The easiest way to stop a war is not to start one in the first place, prevention being better than cure.

3.6 War, what is it good for?

The deadliest military conflict in human history, World War II, lasted six years, and claimed over 10 million lives per year. Annual global deaths from cancer, currently projected to be 14 million, have surpassed that and are on an upward trend. Following the 9/11 attacks on the United States, on September 20, 2001, a mere nine days later, President George Bush declared a "war on terror." What does war create? Having served its intended purpose, the term "war on terror," dutifully echoed by politicians and reported in the media is now passé. Of course, one man's terrorist is another man's freedom fighter. Nelson Mandela was a black man fighting to liberate his country from the repressive forces of an evil apartheid regime that served the interest of a white minority. He was one of the greatest, farsighted, and courageous leaders the world has ever seen, yet he was branded a terrorist by the United States, a country that ironically prides itself on freedom and democracy. As recently as 2008, he was still on the U.S. terrorism watch list, having successfully fought to liberate his country and having served as president of a democratic South Africa from 1994–1999.

Immediately prior to the declaration of war on cancer by President Nixon, war also was declared on narcotics. With the benefit of hindsight, and given the outcome of all three declarations so far, it now seems a tad unwise to declare war against:

- An affliction of which little was known
- An emotional state of extreme fear
- An addictive habit shared by the rich and the foolish

Whatever became of the wars in history when men were men and women were beside themselves? What would the mighty Genghis Khan, conqueror and ruler of the largest land empire in history, make of the notion of battles against such enemies as an affliction, an emotion, and an addiction? In the event of victory, who does one rape, and where exactly does one pillage?

Despite outrageous sums of money spent, and a vast number of man-hours dedicated to wars against cancer and terror, we are far from winning either for many strangely similar reasons:

- Our methods create the very thing we are fighting.
- Failure to appreciate the complexity and enormity of the task at the onset.
- Underestimation of the resourcefulness and obstinacy of the enemy.
- Failure to identify root causes of the problem early on, and address strategy and resources accordingly.
- Some of those engaged in the fight have their own agenda.
- We are attacking an amorphous enemy with unpredictable consequences.
- A lack of focus on prevention.

Traditional ways of treating cancer such as chemotherapy and radiation create cancer. The way cancers are formed is an order of magnitude more complex than originally thought. We don't have a full understanding of which combinations of mutations cause which cancers, how cancers spread, or why some patients respond well to therapy while others don't. We also are fairly clueless about how to stop relapses from happening. Although an increasing number of new drugs are coming onto the market, treatment by surgery, chemotherapy, and radiation therapy are still our main methods of choice. Nevertheless, the hunt for the next blockbuster drug that will save millions of lives and make vast sums of money is on.

Of course, war and cancer are big businesses. When you're the biggest arms dealer around, embracing wars and destabilizing vast regions of the planet are lucrative strategies. Gangsters say, "It's not personal, it's business." Politicians and big business aren't so honest. Like the arms and pharmaceutical industries, government policies are guided by profit and not necessarily by humanitarian values. If we can eradicate smallpox and keep other diseases at bay, with all the technology and wealth of scientific knowledge at our disposal, why are we struggling against cancer? What do we know of it, and how soon can we cure it?

In traditional warfare, the enemy wears a different uniform and invades from outside. Cancer cells wear the same uniform and spread from within. They build up their proliferative and invasive capability in a progressive, patient manner over decades. Unlike viruses and bacteria, no two cancers are alike. This means that a drug that is effective against a form of cancer in one patient may not necessarily be efficacious against the same form in another patient. We face other challenges. For example, how does one create a vaccine to protect against cancer when each form is different and there are hundreds of them.

The majority of anticancer drugs in use today were developed decades ago when we didn't have a clear understanding of what caused the disease. With the sequencing of the human genome, and the rapid and ongoing development of gene sequencing techniques, we can appreciate for the first time the true genetic complexities of the disease.

3.7 The human genome

The Human Genome Project was conceived with the aim of sequencing the human genome and making the results freely available as a resource to researchers worldwide. The project was launched in 1990 with the creation of genome centers in the United States, United Kingdom, France, and Japan. Germany and China came onboard later. After much painstaking research at a cost of three billion dollars, the first draft of the human genome was published in February 2001 about 88% complete. The title of one of the most significant scientific publications in our history was *Initial Sequencing and Analysis of the Human Genome* by Eric S. Lander and another 247 authors. The researchers noted the output of the project would have "profound long-term consequences for medicine, leading to the elucidation of the underlying molecular mechanisms of disease and thereby facilitating the design in many cases of rational diagnostics and therapeutics targeted at those mechanisms." It also was noted that published genome data would permit the comparison of conserved genome segments of different human populations, as well as different species, to provide an understanding of our evolutionary ancestry. A follow-up paper of the whole genome was published in April 2003. Fulfilling the full promise of the Human Genome Project is now in the hands of tens of thousands of scientists around the world, in academia and industry.

From an initial analysis of the first draft, it was suggested the human genome was composed of 30,000–40,000 protein-coding genes. This figure excludes the thousands of genes that code for RNA molecules as their ultimate product, which also carry out biological functions. The figure for the number of genes has subsequently been revised down. It now stands at 19,000. One of the biggest surprises of the human genome projects was only 2% of our DNA codes for proteins. What does the other 98% do? We now understand some of regions that don't code for proteins serve regulatory functions, some code for RNA molecules, and some may be defunct genes, which still leaves a lot of unaccounted for DNA. Among the many highlights of the first draft, surprisingly it was noted that hundreds of human genes appear to have resulted from horizontal transfer from bacteria at some point in the vertebrate lineage.

The wealth of data and greater understanding generated by the Human Genome Project is supplementing new areas of research. It allows scientists to develop a better understanding of genetic disorders so that new ways of diagnosing and treating them can be constructed. Other major genome projects specifically aimed at cancer are under way. These are generating massive amounts of genetic data that has supplemented the nascent field of bioinformatics.

Big data is a term used by different disciplines to refer to data that has potential value, but is so large in volume, and growing at such a rate that organizations are unable to comfortably exploit it using available off-the-shelf technology. Typically, the volume of data is too much, too complex, or is accumulating faster than normal processing capacity. It may be structured, unstructured, or both. When applied to cancer research, big data encompasses gene expression levels, microRNA expression levels, clinical data, and details of genetic aberrations such as simple somatic mutations, copy number mutations, chromosome translocations, and DNA methylation.

The nature, integrity, and consistency of data from different sources are paramount to their usefulness in the elucidation of the genetic causes of different types of cancer. Important considerations in this regard are:

- The type of clinical data collected
- The volume of data provided
- The correct classification of the types and subtypes of cancer

- The use of unique identifiers for parameters such as patients, tumors, samples, genes, and gene transcripts
- The identification and notation of mutations
- The avoidance of missing data or duplication of data

Large-scale global projects that generate and provide access to big data are the International Cancer Genome Consortium (ICGC), The Cancer Genome Atlas (TCGA), the Catalogue of Somatic Mutations in Cancer (COSMIC), and more recently, the 100,000 Genomes Project. In 2006, TCGA began with the aim of cataloguing the genetic mutations responsible for cancer using genome sequencing. TCGA collaborators are engaged in mapping key genomic events for 33 types of cancer. They provide data online for 11,000 patients. The ICGC was launched in 2008 to coordinate genome studies in tumors for 50 cancer types or subtypes. They provide whole genome data on 25,000 tumors arising from the collaborative efforts of 17 countries. The Welcome Trust Sanger Institute maintains curated genomic data from published papers, the ICGC, and TCGA. The COSMIC dataset provides mutation and other data for over one million tumors. The ICGC, TCGA and COSMIC all provide online access to their data via search interfaces, as well as the means of downloading data for local manipulation and analysis.

The 100,000 Genomics Project was launched in 2012 by Genomics England, a company wholly owned and funded by the U.K. Department of Health. Its aims are to collaborate with the National Health Service (NHS) to sequence 100,000 genomes of patients suffering from genetic diseases and cancer in order to:

- Create an ethical and transparent program based on consent
- Bring benefit to patients and set up a genomic medicine service for the NHS
- Enable new scientific discovery and medical insights
- Kick-start the development of a U.K. genomics industry

3.8 The road less travelled

Since 2001, the pace of advance has been swift. Today a human genome can be sequenced at a fraction of the cost and in a fraction of the time, compared to a few years ago. We are at a fork in the road. To the left, is the road well-travelled along the path of which lie surgery, radiation treatment, and chemotherapy. To the right, is the road less travelled that leads to prevention and personalized treatment.

So far, the pharmaceutical industry has depended on a limited number of targets for the development of drugs. It is estimated that virtually all drugs on the market have been aimed at 483 targets. The identification of all the genes of the human genome widens the scope of potential drug targets to thousands. This prospect provides a boost to genomic research in the pharmaceutical industry. The development of new drugs to launch attacks against vulnerable targets peculiar to the form of cancer under treatment is well under way. Major factors are cost, time to market, and market size. It takes 10–15 years to complete all three phases of clinical trials before licensing a drug, at a cost exceeding US$1 billion. Drug companies understandably need to be circumspect in their choice of which cancers to target, and which candidate drugs to develop.

We can safely conclude that we have not won the wars on narcotics, terror or cancer. The war on narcotics can be won tomorrow if consumers stopped buying the evil stuff. We are the ones fueling it. Perhaps some of the money spent on fighting wars would be better spent on prevention. We could, for example, put more effort into educating our kids and making sure they listen. No matter how you spin it, the notion of winning a war on terror by creating the very thing you're fighting does not make sense. If the aim of the enemy is to create terror, you are giving him what he set out to achieve. He has therefore successfully managed to recruit you as an unwitting accomplice. Welcome to my world.

> *If all you have is a hammer, everything*
> *looks like a nail.*
>
> Abraham Maslow

We should be able to win many battles against cancer, and ultimately the war. We are past the first stage of learning, unconscious incompetence, and are now at the entrance hall of conscious incompetence. Ahead of us are conscious competence and, ultimately, unconscious competence.

4
Tumor Growth

A cancer stem cell is an abnormal cell that divides in an uncontrolled manner, giving rise to other abnormal cells that collectively form a tumor. For a normal cell to transform into an abnormal cell, there are two minimal capabilities it needs to have or acquire. First, it must be able to divide, and secondly, it must be able to do so faster than surrounding cells. All tumor cells are descendants of such a rogue cell. From this innocuous beginning a colony grows, acquires new survival skills, and thrives to the extent that it threatens the well-being and very survival of its host. How does all this come about?

To fully transform into a cancer cell, a normal cell needs to overcome many natural obstacles that life forms have evolved to keep uncontrolled cell division in check. The drivers of the stepwise transformation process are random modifications to DNA that by chance, confer growth advantages over other surrounding cells. Cells thus endowed are preferentially selected over time through a Darwinian process, from which a tumor slowly develops. The time to acquire the necessary changes may span decades, but under certain conditions a perfect storm may be formed that accelerates the process. On a daily and unrelenting basis damage to DNA is caused by:

- Errors in its duplication when cells divide
- Random attacks by environmental agents such as carcinogens in smoke, UV light, and x-rays
- Attacks by reactive chemical agents that are byproducts of natural metabolic processes in the body

No genetic mutation on its own is sufficient to overcome all barriers for the full transformation of a normal cell into a cancer cell. We don't know the exact number required. It varies with the type of cell and the type of cancer. It is clear that several biological pathways that coordinate and drive cell division need to be deregulated, as well as other pathways that support them. Pathways are rewired by random mutations that happen to serve the cancer cells' agenda. The agents of which are changes in protein function, brought about by DNA mutations that alter their normal function.

4.1 The hallmarks of cancer

A list of recognizable barriers to malignant transformation has been constructed, referred to as the "hallmarks of cancer." It was first published by

U.S. scientists Douglas Hanahan and Robert Weinberg in 2000 and later revised in 2011. These include:

1. Sustainment of proliferative signaling
2. Evasion of growth suppressors
3. Resistance to cell death
4. Enabling of replicative immortality
5. Inducement of angiogenesis
6. Activation of tissue invasion and metastasis
7. Reprogramming of energy metabolism
8. Evasion of destruction by the immune system

Other noted contributory factors were genetic instability, inflammation, and the recruitment of normal cells to assist transformation.

In summary, cancer cells organize the signaling of their own proliferation. They defuse cell death and senescence. As they grow, they initiate construction of their own network of blood vessels. Not satisfied with such prowess at their primary site, they invade surrounding tissues, and take a trip down river, to alight on distant sites and start new colonies. Somewhere along the way they conjure up replicative immortality, showing complete contempt of protocol and pertinent behavior. Wherein among such audacious acts are there calls to arms by our elite immune system?

> *... Bring me my chariot of fire!*
> *I will not cease from mental fight,*
> *Nor shall my sword sleep in my hand ...*
>
> "Jerusalem" by William Blake

4.1.1 Sustainment of proliferative signaling

Every day millions of cells in our bodies die and millions of new cells are created to replace them. The job of producing new cells is carried out by a small number of select stem cells. The majority of cells in an adult human are fully differentiated and do not have the ability to divide. Those that can replicate only start to divide when they receive signals from other cells telling them to commence. Molecules that command other cells to proliferate are collectively referred to as growth factors. These are small proteins, some act upon a wide variety of different cell types, and some are more specific. For example, epidermal growth factor initiates proliferation of mesenchymal cells, glial cells of the brain, and epithelial cells, while platelet-derived growth factor initiates proliferation of connective tissue, glial cells of the brain, and smooth muscle cells.

Each cell type responds to its own unique array of growth factor, which binds to receptor molecules on its outer membrane. After binding, a chain of events is initiated inside that culminates in cell division. Once the required number of cells is produced, cell division is switched off. The maintenance of a fine balance between cell attrition and cell renewal is essential for the upkeep of tissue size and function. Corruption of the signaling mechanism tips the balance in favor of cell renewal, leading to uncontrolled cell division. This happens when a switch that initiates cells division is left in an on state for longer than normal. The avoidance of cell death also is a contributory factor. There are numerous ways in which these can be accomplished, which will be presented in later chapters.

4.1.2 Evasion of growth suppressors

Once a dividing cell has fulfilled its obligation and supplied new cells, as instructed by neighboring cells, further cell division ceases. Proteins that switch off cell division are tumor suppressors, and genes that code for them are tumor suppressor genes. Mutations of these genes that inactivate them, or inhibit their expression, prevent them from halting cell division. This, in turn, leads to uncontrolled cell division, the principle hallmark of cancer. Normal cells with aspirations of becoming cancerous need to evolve ways of taking out growth suppressors.

In addition to releasing growth factors telling other cells to divide, some cells release growth suppressors that tell them not to divide. Responding cells balance incoming pro-growth and anti-growth signals and reach a decision on whether to divide or not. In part, this is facilitated by pRB, a protein that is present in many cells. It's a key player in the weighing up of signals that originate from outside the cell and inside the cell, and can put a halt to cell division based on outcome. You may recall that retinoblastomas are caused by a deficiency of the RB1 gene. In many cancers pRB is commonly inactivated.

4.1.3 Resistance to cell death

A growth advantage may not be acquired only by an increase in cell division, it also may be achieved by a decrease in cell death. When the DNA of a cell becomes damaged beyond a certain point:

- The risk of it transforming into a tumor cell increases substantially.
- Its ability to serve as a template to produce proteins is impaired.
- Its capacity to serve its normal function is diminished.
- Cell death is triggered by a natural process referred to as apoptosis.

A major barrier to uncontrolled cell division is triggering the death of cells with damaged DNA. There are mechanisms inside cells that continually monitor the integrity of DNA. Where possible, repairs are carried out, but failing that, once damage has accumulated beyond a certain point, cell death is triggered by apoptosis. Transforming cells need to find ways of cutting the wires that set off alarms that trigger cell death. This can be readily achieved by the inactivation of sensor proteins by mutations, a reduction in their levels of expression, or deactivating a protein named p53. Studies have shown the protein mdm2 negatively regulates p53 activity through the induction of its degradation. The role of p53 is illustrated in Figure 4.1.

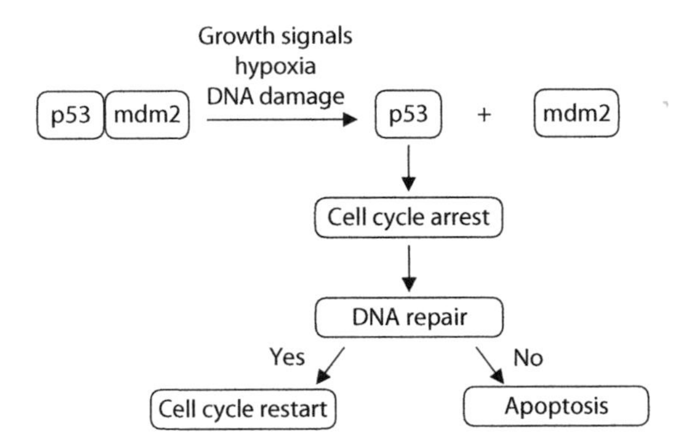

Figure 4.1 Role of p53 in the regulation of DNA repair.

Commonly referred to as "the guardian of the genome," p53 plays a key role in regulating cell division and preventing tumor formation. It is coded for by the tumor protein 53 (TP53) gene. In the cell nucleus, p53 binds directly to DNA, which makes it a transcription factor. It responds to inputs from other proteins that act as sensors of stress inside a cell, such as DNA damage, as well as low levels of oxygen and growth-promoting signals. When it receives a signal that the DNA of a cell is damaged, p53 responds by activating repair enzymes to fix it. If they are unable to do so within a limited timeframe, cell division ceases, and the cell is instructed to commit suicide. Thus, under circumstances of overwhelming damage to cellular subsystems, p53 triggers cell death. By the elimination of cells with damaged DNA, p53 suppresses the development of tumors. The inactivation of p53 is common in a high percentage of tumors.

4.1.4 Enabling of replicative immortality

There is an inbuilt counter in cells that ticks every time they divide. After 40–60 ticks, cells enter a state of senescence, and division is no longer possible. The accumulation of senescent cells is a phenomenon associated with aging. It puts a brake on tumor growth, which cancer cells need to find a way around. They can do this by switching on telomerase, an enzyme that resets the inbuilt counter, which bestows the gift of immortality. Telomerase is inactive in normal differentiated cells and some stem cells, which makes them mortal. Research has shown it is active in 80% of cancers and some stem cells, which gives them a counter without limit. Telomerase activity is primarily switched on by mutations in its promoter region that cause an increase in its levels of expression. The promoter of a gene is a small section of DNA, upstream of the gene, on its chromosome, to which the enzyme DNA polymerase binds when it is being expressed.

4.1.5 Inducement of angiogenesis

Angiogenesis is the physiological process by which new blood vessels are formed through the extension of pre-existing vessel networks. Its purpose is to ensure cells are serviced with an adequate supply of blood, so that they can receive oxygen and nutrients, and get rid of waste products such as carbon dioxide. Our blood vessels are lined with a layer of endothelial cells, through which various molecules are exchanged between the blood and surrounding tissues. As children grow, angiogenesis is carefully coordinated with cell growth, so that each cell receives an adequate supply of blood, without which they would die. Chemical messages are passed between cells that initiate and coordinate the development of new blood vessels when and where needed. Once growth has ceased, further growth of blood vessels is no longer required, and angiogenesis is switched off.

In adults, whose blood networks are fully formed, cells created by a growing tumor become further and further away from existing blood vessels, and become increasingly starved of food and oxygen. Beyond a certain distance, cell growth is not sustainable and further growth ceases. Thus, tumor growth is restricted to a few millimeters. To overcome the barrier imposed by a lack of nutrients and oxygen, the existing network of blood vessels needs to be extended. Cancer cells find ways of switching on angiogenesis or inducing other cells to do so. It's possible that the same system employed when angiogenesis occurs under normal conditions is activated by the release of growth factors by tumor cells, or by surrounding cells under instruction from tumor cells.

With the arrival of appropriate signals, blood vessel epithelial cells commence division to form new branches that grow outwards and extend into tumors. As cells are tightly adhered to each other and to the extracellular matrix, space needs to be created for vessels. The area between cells is filled with proteins and carbohydrates that form the extracellular matrix, which along with cell-to-cell adhesion, keep cells tightly bound together. For angiogenesis to occur, the extracellular matrix needs to be loosened by digestive enzymes to make room for growing blood vessels. The extension of blood vessels into tumors is not as well controlled or constructed as it is under normal growth early in life. Some vessels are leaky and tumor cells may receive less than an adequate supply of nutrients and oxygen. Therefore, tumors can have dead cells on the inside.

4.1.6 Activation of tissue invasion and metastasis

Invasion occurs when tumor cells breach the barrier that separates them from a different tissue type and grow into its space. Tumor cells in the process of invading surrounding tissues have demonstrated migration behavior like that used by normal cells during natural physiological processes, such as growth and wound healing. It's quite possible that migrating malignant cells activate these mechanisms by switching on the expression of one or more sets of genes required for invasion. This would include a reduction of cell adhesion, the ability to clear a path through the tightly bound extracellular matrix, and the attainment of the ability to move. On completion of normal biological processes, for example growth and wound healing, these activities eventually switch off. Unfortunately, the signaling machinery in tumor cells is disrupted, and stop signals are either not sent out or are ignored. This promotes the progression and spread of tumors.

Metastasis occurs when cancer cells separate away from their primary site and start new colonies at distant secondary sites. To accomplish this, cancer cells need to:

- Separate from the colony of their primary site
- Make their way into lymphatic or blood vessels
- Circulate around the body
- Make their out of lymphatic or blood vessels
- Alight at secondary sites and commence growth

It seems incredible that cancer cells can execute all of these steps that normal cells are unable to. Our understanding of the very complex metastasis cascade is quite limited. Given it's the most life-threatening event for patients with cancer, this needs to be addressed. Metastasis is covered in greater detail in Chapter 16.

4.1.7 Reprogramming of energy metabolism

The reprogramming of energy metabolism refers to alterations in cellular metabolic pathways to serve the specific needs and environments of a growing colony of tumor cells. For example, cells further away from blood vessels adapt to metabolize glucose in the presence of low levels of oxygen.

Cancer cells need an abundance of energy and raw materials to divide. The main challenges to cell division are the duplication of the 3.2 billion base pairs that make up the DNA of the human genome, the synthesis of proteins, and the synthesis of lipid molecules for the outer cell membrane and the membranes of internal organelles. Raw materials for these tasks are provided by breaking down carbohydrates, proteins, fats, and DNA from the diet into smaller molecules, which are delivered to cells around the body by the blood circulatory system. Cancer cells compete with other cells for these nutrients.

Cells obtain most of their energy from metabolizing fat molecules and simple carbohydrates, such as glucose and fructose. These nutrients undergo controlled combustion inside cells during which energy is produced and stored in the form of adenosine triphosphate (ATP), the end products being carbon dioxide and water. The full breakdown of one molecule of glucose produces 36 molecules of ATP. This process is essential for sustaining all forms of mammalian life. Molecules of ATP are rich in energy that is released when its bonds are broken to drive biological reactions and processes forward.

To produce energy, one molecule of glucose is first broken down to two molecules of pyruvate in the cytoplasm of cells by a process known as glycolysis. It does not require oxygen and yields two molecules of ATP. As shown in Figure 4.2, the production of the other 34 molecules of ATP is carried out inside small organelles known as mitochondria. These are the powerhouses of cells into which pyruvate is shipped in and combusted with oxygen to produce ATP, carbon dioxide, and water. This oxygen requiring process is referred to as oxidative phosphorylation.

Oxidative phosphorylation is much more efficient than glycolysis for energy production, yet it is common in tumors that glucose is preferentially metabolized in the cytoplasm by glycolysis, even in the presence of oxygen and fully functioning mitochondria. This paradox, known as the "Warburg Effect" after the German scientist who discovered it in the 1930s, has been studied extensively, with a view to answering the question: "What benefits do tumor cells derive from metabolizing glucose by glycolysis which produces 18 times less energy than oxidative phosphorylation?" Despite several explanations, we are not certain about the real advantage. It is currently an enigma wrapped in a mystery. By stopping short at glycolysis and boycotting oxidative phosphorylation, the Warburg Effect shows cancer cells can reprogram their metabolism. It

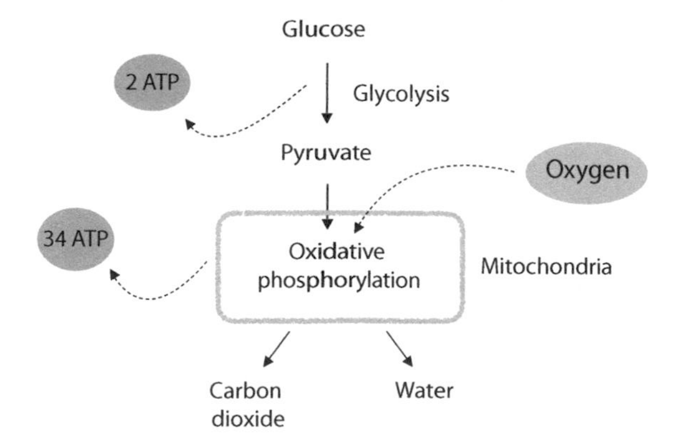

Figure 4.2 Some cancer cells produce energy by glycolysis instead of oxidative phosphorylation.

demonstrated for the first time that the metabolism of tumor cells can be different to that of normal cells.

Bases for the synthesis of DNA are provided in two different ways. They can be salvaged as whole molecules from the DNA of ingested food, or they can be synthesized de novo from the amino acids glutamine, aspartate, glycine, and serine. There are some cells in our bodies that are more dependent on the salvage pathway than others, such as the immune system. This aspect of the metabolism of cancer cells has received little attention, although there are chemotherapeutic drugs that target enzymes involved in the synthesis of bases.

4.1.8 Evasion of destruction by the immune system

The importance of the immune system in protecting us from infections by pathogens such as viruses, bacteria, and parasitic worms is well established. The case for protection against cancer is not so clear-cut. Definitive evidence in support of immunological surveillance keeping cancer development in check is lacking, but there is a considerable amount of circumstantial evidence. For example, immunodeficient humans are associated with an increased risk of developing cancer, as are patients treated with immunosuppressive drugs.

The mere fact that cancer is so prevalent, and that metastasis is so common, suggests that the immune system is only partially effective at preventing it. However, it is difficult to tell what the situation would be without its protection, whatever that may be. Billions of cells, particularly those exposed to environmental carcinogens undergo mutations daily, which accumulate over time. These are either subjected to DNA repair, or failing that, death by apoptosis. The question arises, are cells with damaged DNA that have managed to side-step apoptosis killed by the immune system? Killer cells are primed to destroy any cell that is outwardly different from normal cells beyond a certain threshold. Key questions are: how much of this happens and what is that threshold? We don't have a clear understanding of either. There also is the added complication that some cancer cells manage to escape death by nullifying killer cells of the immune system, or by evading detection. This topic is covered further in later chapters.

4.2 Differentiated cells

Cells specialized for the purpose of division are broadly referred to as stem cells, while those that are specialized to contribute to the function of a particular tissue type are classed as differentiated cells. Based on their potency to spawn different cell types, cells may be organized in a hierarchical fashion. Those with the broadest capability sit at the top of the hierarchy, while differentiated, which do not divide, sit at the bottom. The majority of cells in the body are fully differentiated. Examples are lung, brain, and kidney cells.

Differentiated cells are the foot soldiers of tissues and organs that get their hands dirty and do the heavy lifting. Their duties are to collaborate and execute the mundane, routine functions of the organ to which they belong. They are specialized in the sense that they are equipped with a unique array of proteins suited to their task. The protein content of a fully differentiated lung cell is different than a brain cell, which in turn is different than a kidney cell.

4.3 Stem cells

A stem cell may be defined as any cell with the ability to go through numerous cycles of cell division while maintaining an undifferentiated state. An adult stem cell can renew itself, and can differentiate to yield some, or all, of the major specialized cell types of its specific tissue or organ. Adult stem cells are present in many organs and tissues such as bone marrow, blood vessels, brain, heart, gut, liver, ovarian epithelium, skeletal muscle, skin, testis, and teeth. They are thought to be located at specific sites of each tissue type, called stem cell niches. Stem cells of the small intestine are in crypts in its lining, protected by layers of mucus. Those of white blood cells are stored in bone marrow. It appears that some pericytes, which are contractile cells that compose the outermost layer of small blood vessels, are stem cells.

Stem cells make up a small fraction of the total number of cells in the human body. Some guess the figure to be less than 1%. It has been estimated that one in every 10,000–15,000 bone marrow cells is a stem cell. In the blood, the proportion falls to one in 100,000 blood cells. The transformation of stem cells into cancer cells is restricted in three ways:

- They are fewer in number.
- Most of them do not divide often.
- Some are stored in protected locations away from carcinogens.

Stem cells are special in the sense that they serve as factories to provide cells for growth during development and for replacement of normal cells once growth has ceased. Stem cells may remain in an inactive state for long periods of time until they are called on to divide. It's conceivable that every organ in our body has a stash of precious stem cells with the capacity to produce new cells. As we get older, some of these become less active, which may be a contributory factor to aging.

A fertilized egg is the ultimate stem cell, giving rise to the other 200 or so different cell types in the body. It can divide indefinitely at the point of conception. And it came to pass in those days, the initial stem cell begat two stem cells, which begat four stem cells, and so on. As these develop into a recognizable embryo, which then develops into a fully-grown person, new generations of stem cells lose their original generic persona and become increasingly specialized. Along the way, the gift of indefinite replication is lost. This is not necessarily a bad thing, as it sets the system default to off, preventing runaway cell division.

In between the fertilized egg and differentiated cells are layers of cells that can divide, which are classified with overzealous terminology on the basis of how committed they are to the cause of producing different types of cells. For the sake of simplicity, terms such as totipotent, pluripotent, multipotent, and unipotent shall be set aside to the acute shadows of academia where they belong, and shall not darken the doorstep of your mind. For our purpose, it is sufficient to consider that unlike embryonic stem cells, adult stem cells are limited to producing a small subgroup of one or more cell types. They can reproduce themselves and give rise to other types of cells. At some point stem cells produce progenitor cells that have a limited capacity for self-renewal and which produce terminally differentiated cells, as shown in Figure 4.3.

Some cells, such as blood precursor cells, can give rise to multiple types of blood cells, but can't reproduce themselves, while nerve stem cells stop dividing after full brain development. Adult stem cells may be inactive for long periods of time, but in some organs, such as the gut and bone marrow, they divide on a regular basis. Adult stem cells are distinct from differentiated cells in four important ways:

- They can divide.
- They are unable to carry out any tissue-specific function.
- They can give rise to normal differentiated tissue cells.
- They are fewer compared to normal cells.

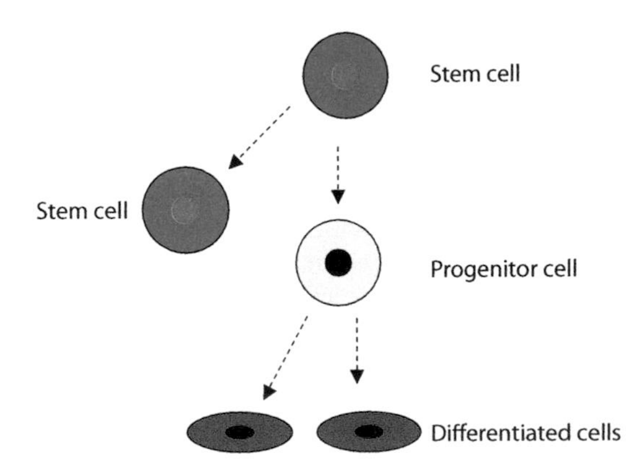

Figure 4.3 Stem cells can divide asymmetrically to produce a new stem cell and a progenitor cell. Progenitor cells divide to produce differentiated cells.

4.4 Malignancy

A colony of growing abnormal cells is called a tumor or a neoplasm. Not all tumors are cancerous. When a tumor acquires the ability to invade surrounding tissues it becomes a cancer, and takes on a new menace. Tumors that breach their site of origin and invade surrounding tissue or spread to other areas, as shown in Figure 4.4, are malignant. Those that do not, such as adenomas, polyps, papillomas, and warts are benign. Benign tumors may cause problems if they grow and press on nerves or blood vessels. A benign tumor in the brain may be life threatening if it remains untreated. In contrast, a malignant tumor that does not spread to other sites may not be as dangerous. Benign tumors may progress to become malignant. To do so they need to acquire new mutations to overcome barriers holding them back. Benign tumors may be hyperplastic or metaplastic. Hyperplastic tissues appear normal except for an excessive number of cells, whereas metaplastic tissues are not normal for the type of tissue.

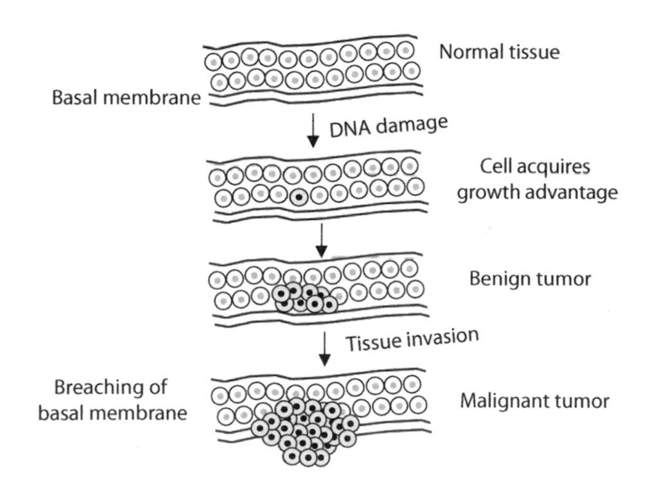

Figure 4.4 Development of cancer from a single cell to a malignant tumor.

4.5 One giant leap

It is possible that the original cell from which a tumor colony develops may originate from a differentiated cell, a progenitor cell, or a stem cell. In theory, any cell with a full complement of DNA can become malignant, but a cell that is already able to divide is already one giant leap ahead. It is not clear whether tumors develop solely from stem cells.

4.6 Tumor growth models

Tumors are complex colonies of different cell types, a significant number of which are unable to divide. Despite intense research efforts, it is not clear exactly how tumors grow. There are two main growth models to consider:

- The cancer stem cell model of growth
- The stochastic model of growth

The models, shown in Figure 4.5, are useful because they provide a focal point and base on which to build and improve. The cancer stem cell model proposes that tumor growth follows the same pattern as normal tissues. It is initiated and driven by stem cells in an organized hierarchical system. According to this model, cancer stem cells with the ability to self-renew indefinitely sit at the top. They produce transit-amplifying cells that can divide a finite number of times. These proliferate and then differentiate to form a third level, which is unable to divide, and therefore does not contribute to tumor growth.

The stochastic model of cancer growth considers all cancer cells to have the same potential to grow and divide, with each cell randomly oscillating between self-renewal and differentiation. The cells of such colonies are not organized in a hierarchical fashion as those of the cancer stem cell model.

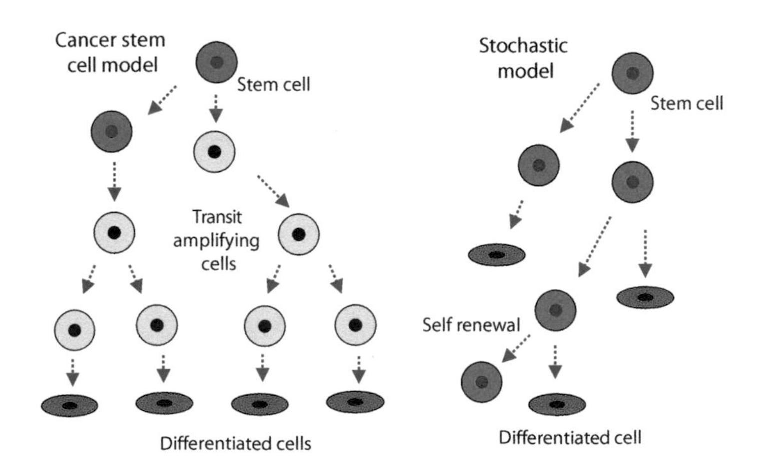

Figure 4.5 Tumor growth models.

Thus, as well as possessing the ability to renew themselves, cancer stem cells are able to generate other types of cells observed in tumors. In the treatment of cancer, it is necessary to kill or remove every single stem cell. There is no definitive proof in favor of either model of cancer growth yet. It's possible that some types of tumors may favor the stochastic model, while others may adopt the cancer stem cell model.

4.7 Branching of tumor colonies

As an initial tumor cell proliferates to form a colony, new variants are formed, each with a modified mutation profile. Over time these evolve into sub-clones, which may, each in turn, form new sub-clones. A growing tumor adopts a tree-like hierarchy where each branch represents a new sub-clone. The older a tumor becomes, the more sub-clones it accumulates, the deeper its branching structure. Cells that, by chance, have mutations that make them fitter for survival in the local environment proliferate at the expense of those less able. All variants are striving to become more cancerous, but some are suppressed by the growth of fitter strains. Therefore, a tumor at an advanced stage is a heterogeneous aggregation of cells dominated by one or more variants of the original cell. The greater the number of sub-clones in a colony of tumor cells, the greater the likelihood of one of them being resistant to treatment. This may be the reason why late-stage tumors are more difficult to treat.

4.8 Stem cell resistance to radiotherapy

The fact that cells of a given tumor are heterogeneous, and constantly are evolving their genetic makeup, is one of the reasons why relapses occur. Treatment regimens that inflict damage on the DNA of tumor cells may by chance promote the development of resistant strains. On initial treatment, sub-clones of cells that are susceptible to a drug die off, but one or more resistant strains may survive to strike back with vengeance in their hearts. In the absence of competing clones that once suppressed their growth, they proliferate with unchallenged vigor (see Figure 4.6). Consequently, drugs need to kill every single cancer stem cell of a tumor.

Radiation works by damaging DNA directly or indirectly to the extent that its integrity is compromised, and cell death is triggered as a natural protective response. Chemicals called reactive oxygen species are generated inside cells by radiation that cause damage to DNA. Cancer stem cells show greater resistance to radiotherapy than normal cells. This may be due to the presence of one or more protective pathways that shields stem cells from DNA damage. Protection occurs in the form of increased expression of proteins that bind to and deactivate reactive oxygen species generated by radiation, hence reducing the damage they cause to DNA. This implies

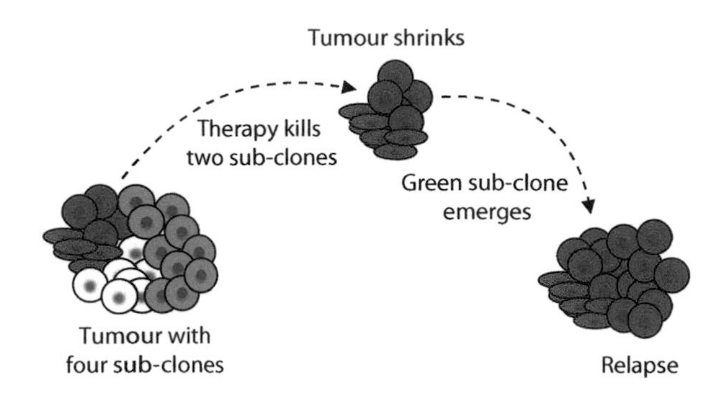

Figure 4.6 Relapse of tumor colony caused by resistant sub-clone of cells.

that the shrinking of tumors upon treatment does not necessarily mean radiation therapy is killing the tumor. Attrition primarily may be due to the death of non-dividing cells, which are more susceptible, leaving cancer stem cells wounded, but very much alive. Experiments have shown that when drugs block protective pathways, stem cells become more susceptible to radiation. Further protection of stem cells to radiotherapy is offered by their higher capacity to repair DNA damage, which has been noted in cancer stem cell populations of glioblastomas, prostate cancers, lung cancers, and breast cancers.

5
Tumor Suppression

We are exposed to millions of pathogens daily through physical contact, ingestion, and inhalation. They pose a clear threat to the lives of all members of society: the young, the old, the good, and the not so good. Protection against the ravages of viruses, bacteria, parasites, fungi, and toxins is provided by an amazing immune system, for which we are truly blessed. It has evolved over billions of years, and though ingenious in design, is exquisitely complex in nature. Were it not up to the task, we would not be around to argue the case.

Infections may take hold of us for a week or so, but most are eliminated by our immune system with brutal efficiency. Many infections are over before they begin. In contrast, the leisurely development of cancer becomes inevitable with age, and once it commences, the body is unable to reverse it without considerable external help. Tumors may disappear for a while, but relapses are frequent. For example, early-stage prostate cancer, deemed to have a high cure rate, has been noted to relapse after five years in 20%–30% of patients. Unlike the case for most infections, there is no assurance that patients declared cancer-free will remain that way. Cancer cells are not the same as normal cells, so where is our much-vaunted immune system when clones of them are spreading throughout our bodies, causing havoc and threatening our very existence?

The idea that the immune system suppresses the development of cancer has circulated for many years. As far back as 1909, the German scientist Paul Ehrlich proposed that were it not for the action of our immune system, the incidence of cancer would be much higher. Is it reasonable to put such a burden of protection against cancer on the immune system? There is much circumstantial evidence to support it. Patients with severely suppressed immune systems, for example those with AIDS and organ transplants, have far higher rates of some cancers. We do have natural systems in place that repress tumorigenesis, such as:

- The inability of differentiated cells to divide
- Cell senescence
- Apoptosis
- Tumor suppressor genes
- DNA repair enzymes

The majority of cells in our bodies are differentiated and incapable of dividing. Only a small number of special stem cells are empowered with this ability. This reduces the risk of uncontrolled cell division. There is a natural limit to the number of times a cell can divide after which it enters a senescent phase. There are biological sensors that detect when the DNA of a cell is damaged beyond repair, and then initiate cell cycle arrest, followed by apoptosis. The elimination of cells with defects in their DNA protects against the development of cancer. There are tumor suppressor genes that put a brake on uncontrolled cell division. They are normal components of any biological process that requires a means of starting and stopping.

As you may recall, if our DNA repair enzymes did their jobs to perfection we would not have cancer, as there would not be any mutations to drive it. The problem with this is, without mutations there would be no evolution. In an ever-changing world, this is not good. We owe a great deal to the heavy hand of evolution, without which we would probably be single-celled organisms swimming around in a murky swamp somewhere. Has evolution compensated for the flaws in our DNA maintenance by providing our immune system with tools to fight cancer?

5.1 The immune burden

Undoubtedly, the primary role of the immune system is to kill invading pathogens that have breached our frontline defenses. Cancer cells may be different from normal cells, but they are not foreign, so is it reasonable to expect our immune system to eliminate them? The only way we would have needed to evolve effective immune defenses against cancer would have been if it threatened our survival as a species. For this to be, cancer would have had to pose a threat to a significant number of us during our most fertile, productive years. Generally, cancer is a disease of the old. The question then is, does the immune system routinely kill incipient cancer cells early in our lives, when its contribution would be of evolutionary significance? If we consider the elimination of pathogens, the answer is a definite yes. There are different ways in which infections contribute to the development of cancer:

- They cause inflammation
- They increase cell division
- Viruses corrupt our DNA
- They release toxins

The association of some viruses with cancer is well established, for example, the human papillomavirus with cervical cancer, hepatitis B and C viruses with liver cancer, and the Epstein-Barr virus with Burkitt's lymphoma and nasopharyngeal cancer. Viruses enter cells, edit our DNA, and highjack our synthetic pathways to turn them into factories that replicate themselves. Crops of virus particles are subsequently released to spread to other cells, killing the host cell in the process. Such sleeping terror cells pose a clear and present threat. While this wanton violation is taking place inside cells, viruses are sheltered from attack by the immune system. As a matter of vital importance infected host cells, already doomed, need to be killed to eliminate them and their viral payload in the process. Because cells infected with viruses are killed by the immune system as a curative measure, infections increase cell division, which increases mutation rates, and in turn increases the risk of developing cancer.

The killing of cancer cells and cells in the process of transformation by the immune system are generally accepted to occur, but the significance is not clear. On the flip side, there is evidence that some immune killer cells not only help to create an environment that sustains the development of cancer, they also protect cancer cells from destruction. What do our genes and biological processes tell us about our immune system's role in the defense of cancer?

5.2 Is evolution bothered?

Over 95% of cancer deaths in the United Kingdom occur past the age of 50. The female of our species becomes infertile around the same age. Not so long ago, living past the age of 65 was rare. It remains true even today, in some parts of the world. In others, it has become quite fashionable. During our 200,000 years of existence was there ever any selective pressure to develop an effective anticancer arm of our immune system? If we weren't fertile long enough for it to matter, and if we didn't live long enough in any case, then probably not. We say, Miss Evolution, "Here is a good anticancer gene, please add it to the gene pool so that the human race may live long and prosper," and she says, "I'm not bothered, do I look bothered?" Evolution is not concerned about how long we live, she only cares about how long we are fertile. Infertile members of our species may help to look after the young, but they do compete for valuable resources without the ability to add to numbers.

5.3 Metastasis

The ability to mutate is a potent contributor to evolution and the long-term survival of species. Cancer cells have the advantage that they mutate faster than our defense system. In some late-stage cancers, malignant cells spread with impunity, via the blood and lymphatic systems to other areas of the body. On their journey, they successfully evade our immune defenses, and subsequently disrupt our finely tuned biological processes to the point of death. The sounding of an apocalyptic horn, and the sight of four horsemen riding by, escapes the attention of the watchful eye of the immune system. Metastasis, which causes 90% of cancer deaths typically becomes deadly decades after the start of primary tumor colonies. Because it generally emerges as a destructive force past our most fertile years, there was never a case for us to evolve ways to deal with it. Metastasis has never threatened our survival as a species, which may explain why we have not evolved to effectively deal with it.

5.4 The immune response

Within the space of 24 hours, a single bacterium, that doubles in number every hour, can produce 20 million copies. When you see someone, and think she is full of vitamins and natural goodness, pathogens are thinking the same thing. Given the fortifying warmth and nourishment of our bodies, and the vast number of pathogens on the loose, time is of the essence in repelling invaders intent on feeding on us. In such times a weak immune system, or one slow to respond, literally may be the death of us.

There are two components to the immune system, the innate immune system, which provides an immediate response, and the adaptive immune system, which, as the name suggests, provides a selective response. In this chapter, we will deal with the suppression of cancer by the innate immune system. The role of the adaptive immune system is covered in the next chapter. We are heavily dependent on the innate immune system to protect us from infection during the first critical hours and days of a new infection.

Our skin and surfaces lining the lungs and the digestive system provide a physical barrier against infection. Tight junctions between the epithelial cells of which they are composed, help to keep pathogens out. In addition, the epithelial surfaces of organs are covered with a layer of mucus that protects against microbial, physical, and chemical attacks. In response to the breaching of these defenses by pathogens, or what it perceives as a threat, the immune system initiates a protective inflammatory response, which develops over time through different phases that encompasses the:

- Destruction of pathogens by killer cells
- Cleaning up of dead cells and cellular debris
- Repair and restoration of tissue to a normal state

There are many different types of white blood cells involved in the destruction of pathogens that will be collectively referred to as killer cells. They form our main defense against pathogens. In addition to the destruction of foreign cells, they can destroy our own cells that have become infected. If killer cells possess the capability to destroy cancer cells, why don't they cure cancer once we have it, as they cure us from invading pathogens?

5.5 The inflammatory response

Inflammation is produced in response to an infection, injury, or what the body interprets as a harmful stimulus. It is often characterized by redness, swelling, warmth, pain, and immobility. Inflammation is an important part of the body's immune response, to heal wounds, and fight off infections. Without it, wounds would fester, and infections would be deadlier.

Cells of the immune system secrete messengers such as cytokines that diffuse and act upon neighboring cells. In response to cytokines, nearby blood vessels put adhesion molecules on the cells of their inner surfaces, to which killer cells circulating in the blood stream, such as neutrophils and natural killer cells bind, and are consequently recruited. These migrate out of blood vessels into infected areas along with fluid. This produces the symptom of swelling, a characteristic of inflammation. Histamine, which is secreted primarily by local mast cells, increases the permeability of blood vessels and blood flow to the affected area. It also enhances the production of mucus to suppress the invasion of pathogens. Mast cells are abundant at sites more likely to suffer injury or infection such as the airways, internal body surfaces, the feet, and blood vessels.

5.6 The innate immune system

The innate immune system encompasses a range of non-specific defense mechanisms that, first, protect against the invasion of pathogens, and second, fight against those that manage to breach frontline obstacles. These mechanisms include physical barriers such as skin, tears, saliva, mucus, and stomach acid, as well as an inflammatory response that uses killer cells to eliminate microorganisms and mop up foreign molecules, such as toxins. Mucus contains substances that kill pathogens or inhibit their growth, which includes a group of antimicrobial molecules referred to as defensins. Tears, mucus, and saliva contain lysozyme, an enzyme that breaks down the cell wall of many bacteria. The various arms of the innate immune system are summarized in Figure 5.1.

As a first call of duty, innate immune cells kill foreign organisms at points of entry, and call upon other circulating immune cells in the blood, such as neutrophils and natural killer cells, to come to the area of infection to help. As a second call of duty, cells of the innate immune system gather intelligence on pathogens, and present it to the adaptive immune system so that a delayed, more specific response can be initiated. Dendritic cells and macrophages carry out this function.

Prominent members of the innate immune system that share the functions of killing pathogens are:

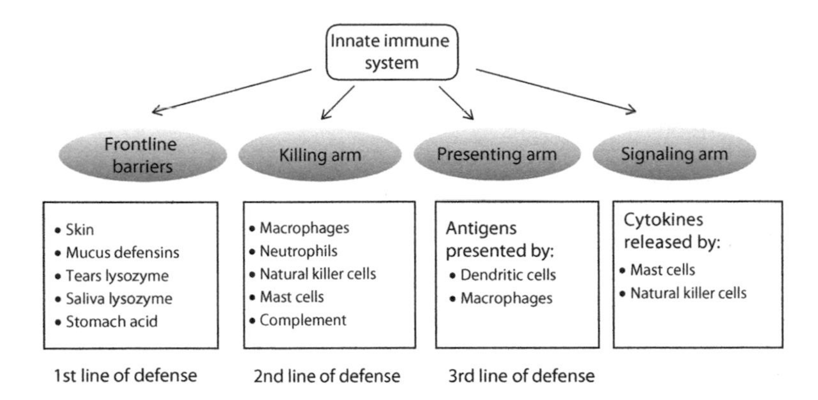

Figure 5.1 The innate immune system.

- Macrophages
- Neutrophils
- Natural killer cells
- Mast cells

5.7 Macrophages

Macrophages start out life as monocytes, which originate from bone marrow stem cells, and differentiate into macrophages when first called into action. They are among the first immune cells to respond to infections because they are especially abundant in areas where pathogens are more likely to gain entry, such as beneath the top layer of the skin, the lungs, the liver, and the digestive system.

As the name implies, macrophages are large, about 16 times bigger than normal cells. Their size gives them the capacity to mop up a high number of invaders. They engulf and digest pathogens by a process known as phagocytosis. This old school technology, employed by creatures such as the humble amoeba for feeding, is used today by elite white blood cells for killing and tidying up. The collective term for immune cells that engulf and destroy pathogens is phagocytes. These are the foot soldiers of the innate immune system.

As foreign cells or foreign particles are engulfed, they are enclosed within membranes to form phagosomes. Digestive enzymes present inside the lysosomes of macrophages are released into the phagosomes, which break down their contents into small fragments, after which they are excreted. Over 60 different enzymes have been found in lysosomes. Short pieces of protein molecules are referred to as peptides. Macrophages journey to lymph nodes, to present peptides of foreign origin to the adaptive immune system, so that a specific response can be launched.

5.8 Neutrophils

Neutrophils are the most abundant white blood cells in circulation, making up to 60% of the total immune cell count. Every day, the human body manufactures and destroys about 100 billion of them. The main function of neutrophils is to kill as many invading microbes as possible. They engulf and destroy bacteria and fungi in a similar manner to macrophages, but are different in the following ways:

- They are not present on site in healthy tissue, but circulate in the blood stream enabling them to respond quickly to calls for help originating from infected sites.
- They possess numerous granules that are specialized lysosomes containing digestive enzymes and antimicrobial peptides.
- They produce extracellular traps that capture and kill microorganisms.
- They die after phagocytosis, forming pus, which is cleaned up by macrophages.
- They do not present fragments of foreign origin to the adaptive immune system.
- They exist for a relatively short period of time.

The lethal molecules of neutrophils not only are destructive to invading microbes, they also are harmful to host cells. Some collateral damage to healthy tissues is inevitable as our bodies fight off infections. The deployment of neutrophil needs to be tightly controlled, and their removal needs to be timely to avoid the accumulation of damage.

5.9 Natural killer cells

As is the case for neutrophils, natural killer cells circulate in the blood ready to be summoned to infection sites in response to cytokine signals. They are different from macrophages and neutrophils in that they are not phagocytes, and they can kill abnormal cells, or normal cells infected with pathogens. Natural killer cells are armed with small granules in their cytoplasm that contain toxic chemicals and proteins, one of which is perforin. Upon release close to a target cell, perforin forms pores in its outer membrane through which other proteins enter. These proteins initiate apoptosis. Importantly, cell death by this means also kills resident viruses, thus preventing their release into the extracellular space. Apoptosis provides a relatively clean form of cell death that does not cause inflammation. In contrast, necrosis caused by trauma promotes inflammation. Natural killer cells also have an immunoregulatory role as they secrete several cytokines. Another important task carried out by natural killer cells is immunosurveillance.

5.10 Mast cells

It is becoming increasingly apparent that mast cells play a role in setting the responses of both innate and adaptive immune systems. They are loaded with an array of biologically active mediators stored in granules that enable them to respond appropriately to various stimuli. Mast cells have the capability to respond directly and indirectly to counter the threat of pathogens. They kill organisms directly by ingestion and chemical attack, and they also secrete antimicrobial peptides. Although their killing function may be important in some infections, their relatively small numbers in frontline tissues suggests greater importance to indirect roles. For instance, mast cells can enhance the recruitment of other immune cells such as natural killer cells, and neutrophils. Immediately upon stimulation, mast cells can release mediators such as histamine, digestive enzymes, and tumor necrosis factor-alpha.

5.11 The complement system

The complement system gets its name from its ability to enhance the binding of antibodies to pathogens, thereby complementing their destruction. An antibody is a protein molecule synthesized for binding to molecules of foreign origin that are collectively termed antigens. The complement system comprises 30 or so enzymes that work together to form an enzyme cascade. Together they indirectly enhance the killing potency of phagocytes by opsonization, and directly destroy microbes by cell lysis. Opsonization is a process by which a pathogen is marked for destruction by phagocytes. It involves the binding of an antibody (an opsonin) to a receptor on the pathogen's outer surface.

The complement system has roots dating back 500 million years, through which a high degree of conservation of structure and function among vertebrates has been maintained, indicative of their importance in defense and survival. Although the complement system forms part of the innate immune system, and is not adaptable, it is called into action by the adaptive immune system. The binding of antibodies to antigens often triggers the complement system through the so-called classical pathway, as shown in Figure 5.2.

Many of the complement system enzymes are present in serum as inactive proteins. The proteins of the enzyme cascade are activated in sequence by the cleavage of peptide fragments by enzymes that are proteases. In essence, protease A cleaves off a peptide from protein B, which converts it into protease B, which then cleaves off a peptide from protein C, which converts it into protease C, and so on. On contact with the surfaces of pathogens, several enzymes are activated in sequence culminating in a cascade of activity that provides three important antimicrobial functions:

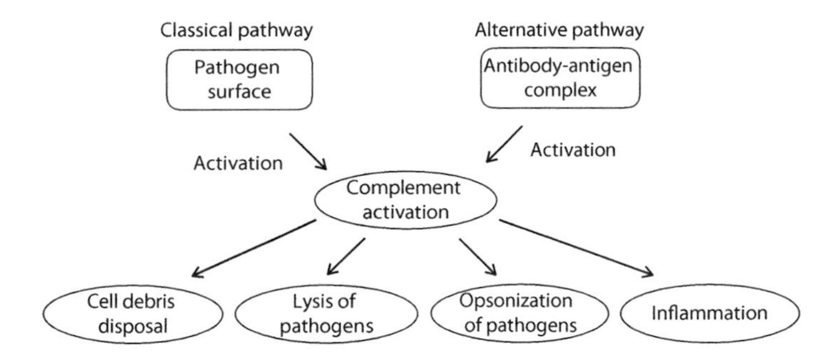

Figure 5.2 The complement system.

1. Activation of inflammation
2. Enhancement of phagocytosis
3. Cell death by lysis

5.11.1 Activation of inflammation

Some of the peptides cleaved off by the action of proteases bind to endothelial cells of blood vessels and lymphocytes, thereby stimulating the release of cytokines, which promote inflammation.

5.11.2 Enhancement of phagocytosis

Complement proteins coat the surface of microbes and increase the likelihood of them being destroyed by phagocytes with matching receptors.

5.11.3 Cell death by lysis

Many proteins of the complement system bind together to form a structure referred to as a membrane attack complex, which can pierce the membrane of pathogens and form pores, thereby leading to cell lysis and death.

5.12 Pathogen recognition

Killer cells of the innate immune system routinely distinguish between cells that are self, and should be left alone, and cells that are foreign, which should be eliminated. To discriminate between friend and foe, the innate immune system has evolved the capability to recognize conserved molecular targets present in commonly encountered pathogens, but are not present in native cells. These targets, referred to as pathogen-associated molecular patterns (PAMPs), are widespread in microorganisms. Innate killer cells have evolved receptors that recognize PAMPs to which they bind. Examples of PAMPs include DNA found in viruses, molecules found in the cell walls of bacteria, and flagellin found in the flagella of bacteria. Killer cells of our innate immune system are armed with receptors on their outer cell membrane that recognize PAMPs. It has been estimated that several hundred exist in vertebrates and are so vital to life that they are encoded in our genes.

Suitable candidate PAMPs need to possess three important characteristics. First, they need to be sufficiently different in structure to human proteins so there is no room for confusion between self and non-self. Second, they need to be essential for the existence of the pathogen to which they belong, ensuring there is little or no scope for structural changes. Third, they need to be an integral part of a broad spectrum of organisms.

This hard-wired, non-specific method of recognition of foreign organisms is built into macrophages, neutrophils, and natural killer cells. It is tried and tested technology inherited and adapted from our ancestors that have been around much longer than we have. The benefit of macrophages, neutrophils, and natural killer cells being preconfigured in this non-specific manner is they can respond immediately to threats. The disadvantage is they can only recognize a limited range of pathogens that by chance have molecular patterns which match their receptors. Any pathogens or cancer cells that fall outside of their scope come under the remit of the adaptive immune system, which can deal with a wider range of pathogens that includes those that slip through the innate net.

5.13 Targeting cancer cells

The fact that they are outwardly different from normal cells makes the identification of viruses and bacteria readily achievable. A normal cell en route to full transformation to the dark side poses a unique problem to our immune system. Molecules on its outer membranes are the same, or nearly the same, as those of normal cells. As such, it may not be recognized as foreign, and therefore may not be killed. The question arises, can the innate technology that recognizes pathogens be used to seek and destroy incipient cancer cells? For this to occur, three events need to take place. First, abnormal proteins need to be produced by transforming cells that are so different from self that they attract the attention of the immune system. Second, these abnormal proteins need to be chopped up into peptides and presented on the outer cell membrane for scrutiny by killer cells. Third, the peptides need to be recognized as foreign by killer cells on surveillance duty.

On a routine basis, large protein molecules are indeed broken down into peptides inside normal cells, which are then attached to proteins, transported to the outer membrane, and put on display to keep the immune system informed of what is going on inside cells. Given the genetic upheaval that accompanies many forms of cancer, the production of abnormal proteins most certainly occurs. However, this chaos tends to occur late on in the development of tumors. To trigger an immune response, proteins do not necessarily have to be drivers of tumorigenesis. Any old protein would suffice.

At the top end of the mutation scale are aberrations involving the duplication of whole chromosomes, loss of whole chromosomes, translocations between chromosomes, frameshift mutations, and deletions and insertions of large chunks of DNA that may cause major changes to proteins. At the other end, are single point mutations that result in the substitution of a single amino acid with a different one. DNA mutations that enhance the function of cancer driving proteins or impair the function of proteins that suppress cancer are drivers of tumor development. All of the different types of mutations that bring about a change in the amino acid sequence of a protein are quite capable of changing it sufficiently enough to affect its function.

There are cancer driver genes that are activated by single point mutations. However, billions of normal cells will have a host of proteins with many mutations that have nothing to do with tumor development. It would be inefficient and possibly disastrous to have them classified as abnormal and killed by the immune system. It is reasonable to assume this does not happen to any great extent. The chances of a single point mutation in a protein a few hundred amino acids in length being present in fragments presented on the outer membrane of a cell in sufficient amounts to trigger an immune response are presumably very small. It has been estimated as many as 10,000 proteins may be presented on the surface of any single cell for scrutiny by the immune system.

5.14 Major histocompatibility complex proteins

For the purpose of presenting protein fragments on the outer membrane of cells, proteins known as major histocompatibility complex (MHC) proteins have evolved. There are two classes, MHC Class I and MHC Class II. MHC Class I proteins are present in every cell of the body, except those without a nucleus. They are used to present fragments of proteins synthesized inside cells, which may be viral peptides or host peptides. MHC Class II proteins are only present in specialized immune cells such as macrophages, dendritic cells, and B-cells of the adaptive immune system. These immune cells ingest foreign cells or molecules of foreign origin, and break them down into smaller fragments, after which they are attached to MHC Class II proteins, and transported to the outer cell membrane as shown in Figure 5.3. Thus, unlike MHC Class I proteins that display proteins that are synthesized inside cells, MHC Class II proteins display proteins that originate from foreign bodies outside cells. These presented peptides provide intelligence to cells of the adaptive immune system.

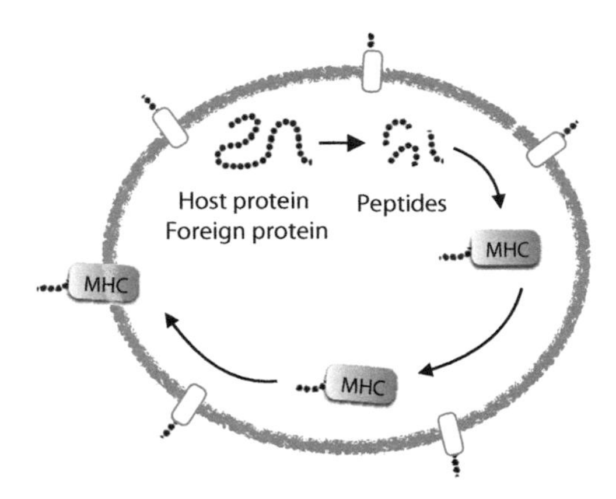

Figure 5.3 Presentation of peptides by MHC proteins.

The term histocompatibility refers to the compatibility of tissues. Major histocompatibility complex proteins are so named because they determine the compatibility of tissues between different individuals. MHC proteins play a major role in the rejection of transplanted organs. Donors with similar MHC proteins are more compatible. Due to the large number of different alleles for them in the wild, on average there is roughly a 10% difference in the makeup of MHC proteins between any two unrelated individuals. This explains why the rejection of donor organs is so common. However, the diversity of MHC proteins serves a protective function, making it more difficult for invading pathogens to elude the host immune systems.

The proteins inside a cell are continually subjected to turnover, some more rapidly than others. Up to 70% of newly synthesized proteins are immediately degraded into peptides because of errors in their amino acid composition or defective folding. A significant proportion of protein fragments presented by MHC Class I molecules are derived from this pool. The presenting of fragments of newly synthesized proteins to immune killer cells enables a quicker response to infections. For example, cells infected with a flu virus become recognizable by the immune system 1.5 hours after infection. Abnormal proteins synthesized by tumor cells also may be broken down and presented to the immune system in a similar manner to those of viral origin.

The need for MHC proteins to bind to a wide variety of peptides is an unusual requirement, as most proteins need to show great specificity. They can do this because their binding pockets are flexible, and the peptides they bind to are small. The smallest peptides recognizable by the immune system are 6–8 amino acids in length. Those attached to MHC Class I proteins are believed to be 9–11 amino acids long compared to 12–25 for MHC Class II proteins. This is because the binding cleft of the MHC Class II protein is

Table 5.1 Presentation of Peptides by MHC Proteins

MHC Protein	MHC Class I	MHC Class II
Present in	All cells with a nucleus	Antigen-presenting cells
Peptide source	Local	Foreign
Size of peptides	9–11	12–25

open-ended, and therefore able to accommodate larger peptide fragments. A summary of the characteristics of MHC Class I and MHC Class II proteins is presented in Table 5.1. It appears that amino acids at positions 4–6 of peptides are more influential at exerting an immune response, as are bulky amino acids.

5.15 Innate immunosurveillance

The unending, daily turnover of billions of cells is critical in the context of tissue integrity, organ function, and the suppression of cancer. Unfit cells and those that have fallen in the line of duty are clinically engulfed and digested by macrophages as part of a cleaning up process. These cells are not necessarily infected with viruses, so how are they targeted for destruction? It turns out there are "eat-me" and "don't-eat-me" proteins that help to regulate cell elimination. For example, calreticulin provides an eat-me signal to macrophages, while cluster of differentiation 47 (CD47) provides a don't-eat-me signal. High levels of eat-me proteins together with low levels of don't-eat-me proteins on the outer membranes of cells together send a strong death wish signal to macrophages on surveillance duty. Calreticulin has been observed on the outer membrane of a diverse array of tumor cell types, which may be a cellular response to transformation. The display of CD47 employed by different cell types of the body that includes tumor cells, prevents spontaneous phagocytic destruction.

As red blood cells become old and worn out by the abrasive forces that accompany circulation, they progressively down-regulate their expression of don't-eat-me signal mediated by CD47 eventually reaching a point where they become prey to macrophages. Primary breast carcinoma cells, that are not invasive, show little CD47 expression, whereas those that have become malignant and metastatic show high levels, presumably to protect themselves from attack by macrophages. The expression of don't-eat-me proteins and the suppression of the synthesis of eat-me proteins are ways in which tumors can evade destruction by phagocytic cells of the innate immune system.

As they circulate around the body, and migrate to and around sites of infection, natural killer cells constantly perform spot checks on other cells they encounter. Normal cells express and display a sufficient amount of MHC Class I complexes on their outer cell membrane to produce a strong inhibitory signal, which protects them from attack. These molecules effectively serve as identification (ID) cards that validate them as native cells of good standing in the community, and are interpreted by natural killer cells that nothing untoward is going on inside. Healthy cells displaying such upstanding traits are left alone by the immune system. However, cells with abnormal peptides on display are served a death sentence.

Many types of human cells display specific proteins on their surface when subjected to stress such as DNA damage, viral infections, or neoplastic transformation. These stress proteins are displayed in addition to antigens. Natural killer cells, possess complementary cell surface receptors to them, and are able to bind to and kill cells to which they belong. Thus, when on immunosurveillance duty, as they scope the outer membranes of cells they encounter, natural killer cells are prompted to use lethal force when they come across:

- An absence or near absence of MHC Class I proteins
- The presence of stress proteins
- The presence of antigen-MHC Class I protein complexes of foreign origin
- The presence of antigen-MHC Class I protein complexes of native origin

Foreign antigens may be synthesized inside cells under viral command, while native antigens may be abnormal peptides produced from faulty gene templates. Some resident viruses and transforming cells hide from the immune system by suppressing the presentation of peptides on the outer cell membrane. Often this is achieved through suppression of the synthesis of MHC Class I proteins. Loss of these is frequently associated with more invasive and metastatic cancers. This has been observed in more than half of advanced breast cancers, a factor that correlates with poor prognosis. In addition, cancer cells that have metastasized to the bone marrow often display little or no MHC Class I proteins. Natural killer cells have the capability to terminate cells deficient in MHC Class I proteins, but it's possible to evade this. One way to do so is to selectively suppress the expression of eat-me proteins or one or more of the six types of MHC Class I molecules that are normally expressed together. This selective suppression may block the presentation of one or more particular antigens, thereby evading the attention of patrolling killer cells.

It seems certain that any normal cell en route to full transformation that synthesizes abnormal proteins will have abnormal peptides on display, and consequently will be killed or tagged for termination by the immune system. We have no idea how many potential cancer cells are eliminated in this way. Candidate executioners of such unhealthy cells are natural killer cells of the innate immune system and cytotoxic T-cells of the adaptive immune system. What contributions do natural killer cells make?

Natural killer cells are each born with the same bunch of receptors on their outer cell membrane, which have evolved to their current composition over millions of years. These are coded for in our DNA. The present mix represents a workable pool of receptors able to recognize a set of conserved structural features across a range of pathogens that has been fine-tuned over time. Human cancer cells are not a species, and do not have a conserved set of proteins to serve as antigens as viruses and bacteria do. Our cancer proteins also were not around during the evolution of our ancestral lineage to bear influence on the current capability of natural killer cells. When hosts die, cancer lineages die with them, unlike pathogens that move from one host to another. We, as a species, have not been in existence long enough for our innate immune system to evolve preconfigured ways to kill our own cancer cells. Therefore, we may safely assume any suppression of cancers by natural killer cells is a minor component of the general scheme of things, and is more likely due to chance rather than design. We need a hero.

6

Adaptive Tumor Suppression

Each day, all creatures great and small are subjected to threats, great and small, from microorganisms that require some form of defense. Primitive multicellular organisms have had to evolve sophisticated innate immune systems to survive. Vertebrates have gone one better with the invention of high-tech adaptive immune systems that are state of the art. Only 1% of animal species on our planet possess such capability. We are thus blessed with two main means of fighting infections that complement each other and together provide an effective force. Our ancient innate component has been conserved during the evolution of mammals and continues to play a critical role in our safe keeping. As it evolved later, and is better suited to respond to a broader range of threats, intuitively one would consider the adaptive immune system to be a more potent weapon in suppressing the evil forces of cancer. Is this so?

The innate immune system evolved to recognize conserved structural features of pathogens and can launch an immediate, non-specific response. In contrast, the adaptive immune system launches a specific response that takes time to develop. It is essentially non-existent at birth, but develops as we become older and are exposed to more pathogens. Three unique features characterize the adaptive immune system:

1. Specificity
2. Diversity
3. Memory

Following infection, the adaptive immune system builds up a response that is specific to a particular pathogen. The amazing thing is that it can do so for virtually every pathogen we are ever likely to encounter in our lifetimes. The implication is that any normal cell, which malfunctions and produces significant amounts of proteins so different from itself that they are in effect foreign, becomes a target. Unlike the innate immune system, the adaptive immune system demonstrates immunological memory. That is, it remembers invading pathogens from past infections and reacts more rapidly to subsequent exposures. In theory, this means if the adaptive immune system killed off a clone of cancer cells, it should not relapse. However, we know relapses are very common.

Among the components of the adaptive system are specialized white blood cells called lymphocytes of which there are two main types, B-cells and T-cells. These are created in the bone marrow by stem cells, and then circulate in the blood. T-cells migrate to the thymus where they reach maturity, hence the name T-cells. Mature B-cells and T-cells are deployed to lymph nodes and other lymphoid organs where they remain on alert to respond to a pathogen invasion. The lymphatic system consists of an extensive network of vessels that connect lymph nodes, the thymus, the spleen, adenoids, and tonsils. Lymph vessels are like arteries and veins that carry vital supplies of blood to all parts of the body. However, they are much finer and carry a colorless liquid called lymph. The primary function of the lymphatic system is to transport lymph, a fluid containing white blood cells, throughout the body. Unlike the blood system, which uses a pump to circulate fluid, movements of the body drive the circulation of the lymphatic system. This is a good reason to exercise regularly. There are hundreds of lymph nodes that serve as traps for pathogens, cancer cells and toxins to facilitate their removal. Thus, swollen lymph nodes may be indicative of an infection or a tumor. Those caused by infections come and go as we become ill and recover. Those caused by cancers are likely to persist.

6.1 The immune response

In response to an invading microorganism we have the capacity to launch three assault waves:

1. An immediate response provided by sentries posted at points of entry
2. A delayed response initiated soon after a breach that is sustained over a period of days involving the recruitment of circulating killer cells to sites of action
3. An adaptive, specific response that takes four to seven days to launch

An immediate response to infections is provided by the innate immune system, the purpose of which is to:

- Kill invading pathogens at points of entry
- Recruit circulating immune cells to sites of infection
- Initiate and support an adaptive response

An immediate response serves to overcome a variety of threats that our defenses are preconfigured to deal with, and to provide intelligence to the adaptive immune system so that it can ramp up production of antibodies and killer cells. Consequently, the adaptive immune response joins the war effort late, as some allies are prone to do. By the time it joins the fight, we may already be very ill.

6.2 Antigens

Foreign bodies are recognized because they come loaded with antigens. These are molecules or parts of molecules that can stimulate an immune response. They are typically of foreign origin, but can also be produced locally in the body. Antigens may present as free-floating molecules, such as toxins, or as parts of larger structures, such as pollen, or as the cell walls of viruses and bacteria. Foreign antigens originate from a wide variety of sources that include pathogens, poisonous creatures, certain proteins in foods, and tissues from other individuals. Normally, the body can distinguish between molecules that are self and those that are non-self, but in the case of autoimmune disorders normal molecules provoke an immune response. Large molecules are common antigens, but small molecules typically are not.

6.3 Antibodies

An antibody is a protein synthesized by plasma cells of the immune system for the purpose of binding to antigens. They are also referred to as immunoglobulins or gamma globulins. The binding of antibodies to the antigens of pathogens impedes their virulence by preventing them from binding to host cells, and tags them for destruction by the immune system. The binding of antibodies to free foreign molecules serves to nullify any potential toxic effects and tags them for removal.

Antibody molecules are Y-shaped. The V-end of the Y attaches to complementary antigens, while the tail end serves as an attachment point to which killer cells can dock, as shown in Figure 6.1. Thus, antibodies form bridges to foreign molecules, pathogens, and normal cells displaying antigens that facilitate attacks by killer cells of the immune system. The binding

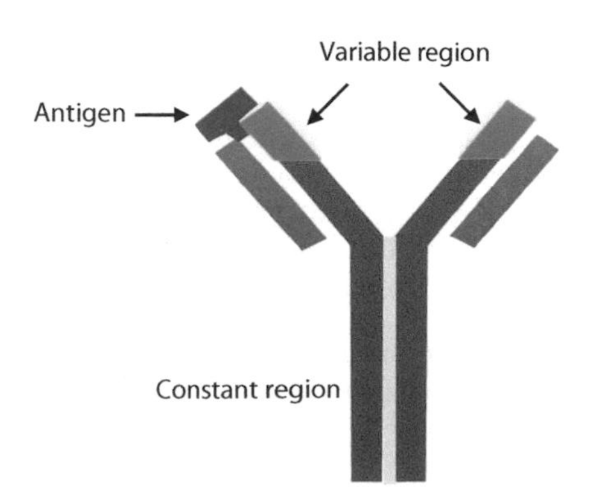

Figure 6.1 Antibody with attached antigen.

of antibodies to pathogens or cancer cells that marks them for destruction may occur in two different ways. First, the formation of antibody–antigen complexes makes it easier for macrophages and neutrophils to engulf the pathogen or cancer cell. Second, it activates complement, the enzyme cascade system that kills target cells by breaking them open.

When an antibody binds to an antigen, it does so to a small accessible segment known as an epitope. In the case of protein antigens, these may be as small as six amino acids. A typical protein, which contains hundreds of amino acids, therefore may have multitude epitopes.

6.4 The problem with epitopes

The problem with stuff is there's so much of it. Epitopes are the same. First, how is an overworked immune system to cater for the billions of different epitopes? Second, how is an immune system to distinguish between epitopes that are self from those that are non-self? Consider, there are 21 different amino acids in man, which means for a small epitope of six amino acids there are over 85 million possible amino acid sequences out of which some will be self and the rest non-self. It's not possible for our genes to code for receptors for even a fraction of these. The immune system has come up with two ingenious solutions to the problem with epitopes. When all else fails try ingenuity.

During the development of immunity, our bodies produce millions of different antibodies each with its own unique V-end, but with the same tail part of the Y, as shown in Figure 6.1. The variable V-part of each antibody is constructed by the random shuffling of the DNA templates used to make it. This structural feature means the antibodies of each person are collectively able to bind to millions of different antigens, while at the same time maintaining the capability to provide docking sites for killer cells. This simple solution provides broad specificity without the need for millions of different genes to code for millions of different antibodies. There are significant differences in the design of the innate and adaptive arms of the immune system that have implications when it comes to cancer suppression. The adaptive immune system with its vast variety of antigens is much more likely to target abnormal cells, including cancer cells. This addresses the first problem with epitopes. The second problem of how to distinguish self from non-self is addressed by the selective rearing of B-cells.

In readiness for the invasion of foreign molecules or pathogens, the body produces billions of different B-cells, each with its own unique antibody. There may be thousands of antibody molecules on the outer membrane of each B-cell, but they are all the same. Early in the development of our immune system, as part of quality control, the production of any B-cell that produces an antibody that binds to self is terminated. We are left with upstanding, patriotic B-cells that only produce antibodies to molecules that are of foreign origin. This means any tumor cells present at birth displaying abnormal proteins may escape attack from the adaptive immune system.

Antibodies patrol the immune landscape in three forms:

- As free-floating molecules in blood plasma
- Attached to the surface of circulating B-cells
- Attached to the surface of cytotoxic T-cells

6.5 Strangers in the night

It is normal for cells to display protein fragments that have been synthesized inside on the outer cell membrane attached to MHC Class I proteins. Tumor cells, or those infected with a pathogen, attract the immune system's attention if these fragments are not self. Like strangers in the night exchanging glances, passing natural killer cells with complimentary receptors, or cytotoxic T-cells with complimentary antibodies, are allured by this provocative act, and bind in polite submission. Seduced, betrayal swiftly follows as circulating killer cells are alerted of this dalliance. In a short space of time, they arrive at the scene sirens blaring, and blue lights flashing to extirpate intruders and infected cells. Thus, the killing of infected cells that are converted factories to produce viruses is a means of keeping infections in check. Free-floating viruses and bacteria loaded with foreign antigens on their surfaces follow the same precipitous fate, as do any tumor cells foolish enough to put abnormal proteins fragments on display. Avoid attentive strangers in the night.

6.6 Dendritic cells

Dendritic cells get their name from their surface projections that resemble the dendrites of neurons, a type of cell used to relay messages in the nervous system. Immature dendritic cells are produced in the bone marrow from where they migrate throughout the body. They remain dormant until they encounter invading pathogens or molecules of foreign origin, which they engulf, process, and present. Dendritic cells are found in most tissues of the body and are particularly abundant in areas where they are more likely to encounter antigens such as the skin, lungs, and gastrointestinal tract. Although macrophages also present antigens, dendritic cells are better at it, and often are described as professional antigen-presenting cells. For this function, dendritic cells have an enhanced capability to uptake antigens and process them for presentation on their outer cell membrane. Macrophages and dendritic cells use both MHC Class I and MHC Class II molecules to present antigens. Other cells of the body use just MHC Class I molecules. Dendritic cells also help to control the function of T-cells and other types of lymphocytes.

6.7 B-cells

B-cells perform several critical roles for the adaptive immune system that include the production of antibodies, presentation of antigens, and production of regulatory cytokines. They bind to soluble antigen molecules present in the extracellular fluid, or on the surface of antigen-presenting cells such as macrophages and dendritic cells. After binding, antigens are engulfed, broken down into fragments, attached to MHC Class II protein molecules, and presented on the outer cell membrane. The binding of an antigen to a B-cell activates it. Helper T-cells attach to such activated B-cells, and stimulate them to divide by secreting chemical messengers. Consequently, the number of B-cells doubles every six hours, so that after a week about 20,000 are generated each able to produce the same antibody. Many of these B-cells then quickly differentiate into plasma cells that are suited to ramp up production of their antibody, producing 2000 per second. This helps to make the immune response specific to an invading pathogen. As a given microorganism has any number of antigens, it's very likely a single infection will trigger ramping up production of many different antibodies.

6.8 Cytotoxic T-cells

Cytotoxic T-cells are like B-cells in that each carries thousands of copies of its own unique antibody. The difference is that unlike B-cells, they are killers that destroy cells infected with viruses. Cytotoxic T-cells use cluster of differentiation 8 (CD8) co-receptors to bind to infected cells and abnormal cells displaying antigens. On binding to their cognate antigens, cytotoxic T-cells become activated and are prompted to multiply by helper T-cells in a similar fashion to B-cells. However, whereas B-cells release their antibodies, T-cells display theirs on the outer cell membrane. In this manner, one or more clones of T-cells, specially armed to bind to and kill a specific pathogen, are produced. This process also takes about a week to complete, and accounts for the delayed response of the adaptive immune system.

Cytotoxic T-cells employ two types of toxic proteins to kill infected cells, perforin, which punches holes in the outer membranes of target cells, and granzymes, which enter through pores, and induce apoptosis. When a cytotoxic T-cell comes across a cell with foreign epitopes on display, its membrane protrusions explore the surface of the target cell. It then binds to it, makes a hole, and injects its payload of poisons. This method provides cytotoxic T-cells with the means to destroy virtually any infected cell with great precision, sparing adjacent normal cells in the process. Such precision is critical in minimizing tissue damage and maximizing the eradication of infected cells. Can this technology be used to launch precisions strikes against cancer cells? Like antibodies on a virus, scientists are on it.

Although they both kill cells infected with viruses, there is an important distinction between the capabilities of natural killer cells of the innate immune system and cytotoxic T-cells of the adaptive immune system. Both can kill cells displaying antigens attached to MHC Class I proteins on their outer membrane, but natural killer cells also kill cells with suppressed levels of MHC Class I proteins. Thus, natural killer cells are primed to eliminate cells that cytotoxic T-cells miss, because they do not display antigens, while cytotoxic T-cells kill cells that natural killer cells may miss because they (natural killer cells) only recognize a limited range of signature pathogen antigens. The arming of natural killer cells of the innate immune system, with a limited number of receptors coded for by genes, means they are less likely to be of significance in eliminating incipient cancer cells than cytotoxic T-cells, which are primed with wide array of protein antibodies. There are billions of different cytotoxic T-cells, each with its own unique antibody receptor. This receptor enables the adaptive immune system to attack virtually any cell displaying molecules that are not self, including incipient tumor cells.

There are a number of factors that favor detection avoidance of incipient tumor cells by the immune system:

- Driver mutations may not produce a change in the structure of protein fragments presented by MHC Class I proteins sufficient enough to class them as non-self.
- Peptides of mutated proteins may not be presented in detectable amounts on the outer membrane, because either they aren't produced in sufficient quantities or they don't bind well to MHC Class I proteins.
- Natural killer cells of the innate immune system that are hard-wired to seek out protein fragments of pathogens are not configured to recognize mutated peptides arising out of self.
- It's not possible to present peptides of proteins, such as tumor suppressor proteins, that contribute to the development of tumors by their absence.
- Proteins that have not mutated, but drive tumor development because they are under expressed, will not be perceived as abnormal, because they are self.
- Proteins that have not mutated, but drive tumor development because they are over expressed, may not necessarily be perceived as abnormal, because they are self.
- Genetic upheaval tends to occur late in the development of tumors, by which time it may be too late.
- Tumors evolve several different means of suppressing an effective immune response.

There is good evidence that some tumors evade the immune system by suppressing the immune response or by using cloaking technology. For example, by switching off immune killer cells or by recruiting normal cells to act as barriers. This introduces the concepts of subterfuge and customization of the tumor microenvironment, turning the hood into a den of inequity.

6.9 Helper T-cells

Helper T-cells are arguably the most important cells of the adaptive immune system, as they prompt:

- B-cells to multiply and secrete antibodies
- Cytotoxic T-cells to multiply and kill infected cells

Helper T-cells themselves are only activated to become effector cells when they encounter antigen-presenting cells. They use cluster of differentiation 4 (CD4) co-receptors on their outer cell membrane to bind to activated B-cells and cytotoxic T-cells in the act of prompting them to multiply. The relevance of helper T-cells to immunity is emphatically demonstrated by AIDS, a condition in which helper T-cells are attacked by the human immunodeficiency virus (HIV) that renders patients unable to defend themselves effectively against pathogens, many of which are normally harmless. The roles of B-cells, cytotoxic T-cells, and helper T-cells of the adaptive immune system are summarized in Figure 6.2.

6.10 Acquired immune deficiency syndrome and cancer

As stated above, helper T-cells use CD4 co-receptors on their outer cell membrane to bind to activated B-cells and cytotoxic T-cells in the act of

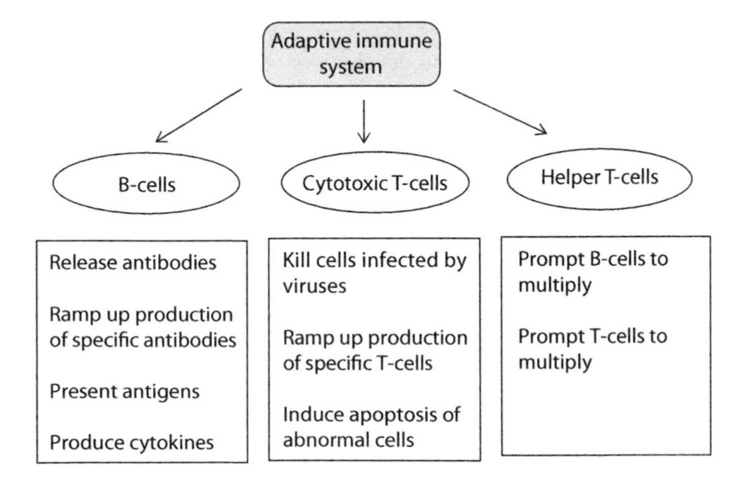

Figure 6.2 The adaptive immune system.

Table 6.1 Presentation of Peptides by MHC Proteins to T-cells

MHC Proteins	MHC Class I	MHC Class II
Present in	All cells with a nucleus	Antigen-presenting cells
Peptide source	Local	Foreign
Peptides presented to	CD8 cytotoxic T-cells natural killer cells	CD4 helper T-cells

prompting them to multiply. In contrast, cytotoxic T-cells use CD8 co-receptors to bind to infected cells and abnormal cells displaying antigens. The presentation of peptides by MHC proteins to the different types of T-cells is summarized in Table 6.1.

HIV selectively infects helper T-cells, because it binds to CD4 co-receptors to enter the cells it infects. For this reason, a CD4 cell count is a lab test used to assess the progression of HIV infections into full-blown AIDS. A reduction in the number of helper T-cells compromises the immune response in patients with AIDS, making them more susceptible to infections and cancer. If the adaptive immune system plays a major role in suppressing cancer, one would expect patients with AIDS to be susceptible to various types of cancers. The profile of cancers that inflict AIDS patients therefore provides an insight into the defense capabilities of the adaptive immune system.

Individuals infected with HIV demonstrate a much higher risk of developing certain types of cancer compared to the general population. According to the U.S. National Cancer Institute, for those infected with HIV, the risk of developing:

- Kaposi's sarcoma is several thousand times greater
- Non-Hodgkin's lymphoma is 70 times greater
- Anal cancer is 25 times greater
- Hodgkin's lymphoma is at least 10 times greater
- Liver cancer is five times greater
- Cervical cancer is five times greater
- Lung cancer is three times greater

Kaposi's sarcoma, non-Hodgkin's lymphoma, and cervical cancer are regarded as AIDS-defining cancers that signal the progression of an HIV infection into AIDS. What do the cancers that AIDS patients suffer from have in common? Many are associated with viral infections, for example:

- Kaposi's sarcoma with human herpes virus
- Non-Hodgkin's lymphoma with the Epstein-Barr virus
- Anal cancer with human papilloma virus

- Hodgkin's lymphoma with the Epstein-Barr virus
- Cervical cancer with the human papillomavirus
- Liver cancer with hepatitis B and C viruses

Infections of the above viruses are more common among those infected with HIV. Some mouth and throat cancers are linked to infection of the human papillomavirus virus, the same virus that causes cervical cancer. In addition, there is a greater prevalence of some well-known risk factors among members of the HIV infected population. For example, lung cancer is associated with smoking, and cancers of the mouth, throat, and liver are associated with heavy alcohol consumption.

According to the U.S. National Cancer Institute , patients infected with HIV share the same risk of breast, colorectal, prostate, and many other common types of cancer as members of the general population. Virally induced tumors occur at much higher rates in immunodeficient hosts in contrast to non-viral tumors, which occur at nearly equal rates. This suggests our adaptive immune system does not offer much by way of protection against three major types of cancer. The most significant protection our immune system offers appears to be via the reduction of viral infections. One could go so far as to contest the generally held belief that our immune system suppresses the progression of cancer by the routine elimination of significant numbers of tumor cells. However, antibodies to tumors have been observed in the blood of patients suffering from different types of cancer and is in itself indicative of some form of immunoresponse. Yet, it is unclear whether such antibodies make an effective contribution to the elimination of tumor cells from the body.

There is significant evidence supporting the immunosurveillance hypothesis derived from observations that cancer cells, like viruses, go to great lengths to protect themselves from the immune system. For example, tumor cells are known to down-regulate the expression of antigens that provoke an immune response. They also suppress the display of the antigen-MHC Class I protein complexes, which are often absent from metastatic cells. Some cancer cells develop the capability to kill immune killer cells, by releasing chemical such as transforming growth factor β (TGF-β) that promotes apoptosis.

6.11 Chronic inflammation

A complete reversal of the concept of the immune system keeping tumorigenesis under control is its apparent complicity in the development of some cancers. A smoking gun is the pernicious presence of long-term inflammation. For example, decades of gallstones and the resulting inflammation are associated with gallbladder cancers. Similarly, liver carcinomas, which are common in some parts of the world, are associated with inflammation that accompanies chronic hepatitis B virus infections. Research has shown that mutations of genes that control inflammation are associated with higher risks of cancer progression.

Short-term inflammation is good, because it shows the immune system is fighting off infections or healing wounds. However, long-term or chronic inflammation is not good. Rudolf Virchow, a German physician, noted a link between inflammation and cancer as early as 1863. The response of the body to a tumor is not unique, and has much in common with inflammation and wound healing. The relationship between chronic inflammation and tumor development has led scientists to compare tumors to chronic wounds. It has been suggested this response is more likely to contribute to tumor growth, progression, and immunosuppression than it is to contribute to the destruction of tumors.

Generally, chronic inflammation is accepted to be a contributory factor to tumor development. It happens when the body launches an inflammatory response to a threat that persists over a long period of time. The threat may be real, such as a virus that won't go away because it's able to avoid being killed by the immune system, or it may not be real, such as rheumatoid arthritis, when the immune system turns against normal cells. Although the exact mechanism is not always understood, chronic inflammation is associated with a variety of habitual and environmental factors, such as smoking, stress, alcohol abuse, obesity, poor diet, a lack of exercise, pollution, and poor oral health. Chronic inflammation is not readily apparent because it does not always present outward symptoms.

7
Surgery

We shall not cease from exploration.
And the end of all our exploring will be to arrive where we started,
and know the place for the first time.

"Little Gidding" by T. S. Eliot

Surgery is the oldest form of cancer treatment. It is a medical procedure inherently associated with risk. A common cause of concern is the possibility of infection, which can become a serious complicating factor. About 60% of cancer patients undergo surgery, often in conjunction with other forms of treatment. Whether it is a viable option or not depends on:

- The type of tumor
- The stage of the tumor
- The risk of causing damage to organs or nerve fibers
- The general health of the patient

Varying combinations of surgery, radiotherapy, and drug treatment have constituted our main attempts at curing cancer, and failing that, at prolonging and improving the quality of life. Surgery is at its most effective when every cell of a tumor is removed, but it is not always possible or safe to do so. We can say a particular treatment regimen that includes surgery will work for a certain percentage of patients, but we are unable to tell with a high degree of certainty how effective it will be in the long run for a particular patient. A lot depends upon the genetic causes of the particular tumor, how advanced it is, the patient's state of health, and the patient's genetic traits.

Cancers that spread from their primary sites to other parts of the body do so via the lymphatic system, the bloodstream, or both. The presence of cancer cells in lymph nodes provides a clear indication of metastasis. Sentinel nodes, which are the first nodes of the lymph circulatory system downstream of a tumor, are among the first areas where cancers migrate. These are examined for the presence of cancer cells to determine the stage of a tumor, and to assess if they need to be removed along, with the primary tumor. Extensive spread to lymph nodes provides strong evidence that cancers are at an advanced stage. Nodes further away from tumors than sentinel nodes are auxiliary nodes. In many surgical operations, some of these also are removed, along with the primary tumor and sentinel nodes as a precautionary measure.

Lymph nodes filter bacteria and other harmful substances from the lymphatic fluid. Lymphatic fluid contains white blood cells that are employed to fight off infections. The removal of lymph nodes increases the length of time spent on the operating table. It also may deprive the body of effective means of draining off lymph fluid causing it to accumulate, a condition known as lymphedema, for which there is no cure. The more lymph nodes removed, the greater the chances of lymphedema happening along with infections, pain, stiffness, and numbness. It must be stressed that removal of lymph nodes does not necessarily prevent the spread of cancers as metastasis also occurs via the bloodstream.

The traditional treatment for breast cancer has been the removal of primary tumors, as well as lymph nodes underneath the arm. A study of 400 breast cancer patients showed that 15% developed lymphedema in an arm after removal of axillary lymph nodes, in marked contrast to 2% who only had sentinel lymph nodes removed.

It has been suggested that many breast cancer patients, particularly those with relatively early-stage cancers, may not need to have their axillary lymph nodes removed to improve their chances of survival. There is evidence to support this. A study of 891 women with breast tumors 5 cm or less in diameter that had spread to their lymph nodes showed those who had sentinel nodes removed had the same five-year survival rate as those who had 10 or more nodes removed. Both groups had surgery to remove their tumors and radiation of their entire breasts.

7.1 Surgery in cancer treatment

Surgery used for cancer treatment can be divided into six types:

1. Curative
2. Preventive
3. Diagnostic
4. Staging
5. Reconstructive
6. Palliative

7.1.1 Curative surgery

Curative surgery involves the cutting out and removal of cancerous tissue. Some normal surrounding tissue, known as clear margin, also may be removed as part of the procedure. Surgeons may use visual aids, such as microscopes, to improve the work accuracy. If it's not possible to remove a whole tumor, a surgeon may opt to reduce its size as much as possible, a process referred to as debulking. Surgery is often used in combination with chemotherapy or radiation treatment to either shrink tumors beforehand, or to increase the chances of killing off any remaining tumor cells afterward. A smaller tumor may increase the effective outcome of chemotherapy or radiation treatment.

During operations surgeons avoid cutting into tumors as much as possible to minimize the risk of inadvertent spreading of tumor cells. Cancer cells that separate away from a tumor are quite likely to end up in a nearby lymph node, which is why they are prime candidates for removal.

> *It's not what you look at that matters, it's what you see.*
>
> Henry David Thoreau

Tissue removed by surgery is examined under a microscope to determine if any further treatment is required and to address relapse risk. Surgery is more likely to be curative against small, early-stage cancers that have not spread to other parts of the body. Therefore, early cancer diagnosis is critically important to the successful use of surgery, or any other form of treatment, for that matter. After a tumor has metastasized, surgery is not curative.

7.1.2 Preventive surgery

Preventive surgery is used to remove tissue that is not cancerous, but is at high risk. For example, the removal of precancerous polyps in the colon before they become malignant. It is well known a woman's risk of developing breast or ovarian cancer is greatly increased if she inherits defective BRCA genes. There are two BRCA genes: BRCA1 (breast cancer 1) and BRCA2 (breast cancer 2). Some ethnic groups are more prone to some forms of cancers than others. For example, faulty BRCA1 genes are present in 1.3% of African American women compared to around 9% of Ashkenazi Jewish women. A woman at a high risk of developing breast cancer may choose to have a prophylactic mastectomy rather than run the risk of dying from the disease.

7.1.3 Diagnostic surgery

Diagnostic surgery involves the removal of a small piece of tissue, or a whole tumor, for examination under a microscope to determine if a growth is malignant or not. Normal cells are uniform and orderly. Cancer cells vary in size and shape, and are noticeably different from normal cells. They often have a large, irregularly shaped nucleus and a smaller amount of cytoplasm. In most normal cells, the nucleus is about one-fifth the size of the cell; in cancer cells, it may occupy most of the cell. Tumor cells often lack some of the functionality of their normal counterparts such as the ability to secrete mucus, or the production of keratin by skin cells.

7.1.4 Staging surgery

Staging surgery is used to assess how much cancer a patient has and where it is located. The primary aims are to determine the size, number, and location of all tumors. It is critically important to establish if a tumor has spread to other areas. This information helps to plan a course of treatment, and to predict the effectiveness of the treatment. It also provides a baseline that can be used to monitor a patient's response to treatment.

One method of assessing if a tumor has spread is to examine lymph nodes to see if there are any abnormalities. Cancer cells that break away from the primary tumors may become trapped in lymph nodes. Imaging techniques such as x-rays, ultrasound, and computed tomography (CT), magnetic resonance imaging (MRI), and positron emission tomography (PET) scans also are used to locate and size tumors. Advances in imaging technologies mean tumors are identified and targeted more accurately for surgery and radiotherapy.

If a lymph node is swollen, an ultrasound may be used to carry out further checks. Swollen lymph nodes also can be caused by infections. If the ultrasound image of a lymph node is abnormal, a fine needle aspiration on the node may be carried out. This involves the withdrawal of a cell sample into a syringe, which is then examined under a microscope. Surgery can be used to remove nodes infected with cancer cells.

A doctor may use an endoscope to examine hollow body cavities and organs such as the lungs, as well as the intestinal or urinary tracts. In addition to allowing the visual inspection of a suspicious area, endoscopes can take tissue samples. The abdominal cavity may be examined under general anesthesia using a laparoscope, with a small incision enabling entry.

Cancers afflicting different patients at the same stage tend to be treated in the same manner. For an early-stage malignant tumor that has not spread, local treatment such as surgery and/or radiotherapy may be sufficient. However, if a cancer has spread, then local treatment will not suffice. A form of treatment that can reach the whole body, such as chemotherapy, hormone therapy, immunotherapy, or another drug treatment is required.

7.1.5 Reconstructive surgery

The aim of reconstructive surgery is to return the body to normal appearance and function following treatment. The most common restorative surgery is breast reconstruction after a mastectomy. Facial reconstruction and testicular implants also are examples of reconstructive surgery.

7.1.6 Palliative surgery

Palliative surgery is used to ease pain, disability, or other complications that arise with advanced cancers. Its aim is to improve the quality of life, not to cure the disease. If a tumor is inaccessible, or if it has invaded surrounding tissue to a large degree, or if it has spread, palliative surgery provides a means of relieving symptoms. For example, if a tumor presses on a nerve, it may be cut back to relieve pain. If one creates a block in the digestive system, it may be cut back to improve function. Surgery also may be used to stop bleeding, or to insert metal rods to protect bones weakened by cancer.

7.2 Surgical Methods

We all are familiar with the method of applying sharp tools, such as scalpels and scissors, to cut out tumors under local or general anesthesia. There are a number of other effective options that can be used to destroy or remove cancerous or precancerous growths. These include:

1. Laparoscopic surgery
2. Robotic surgery
3. Mohs surgery
4. Cryosurgery
5. Electrosurgery
6. Laser treatment
7. Laser-Induced Interstitial Thermotherapy (LITT)
8. Photodynamic therapy
9. Natural orifice surgery

7.2.1 Laparoscopic surgery

To perform routine abdominal surgery, surgeons need access via an opening large enough to provide adequate visibility, with sufficient room to maneuver and operate surgical instruments. To provide this, a single abdominal incision is made that may be 15–30 cm long, depending on the size of the patient and type of operation being conducted. A significant amount of postoperative pain is associated with such large incisions, which prolong time spent in hospitals and slow overall recovery.

Minimally invasive surgery is an innovation that avoids use of a single large incision that is standard in traditional open surgery. Laparoscopic surgery is a technique employed to operate on patients via several small incisions, about 1.5 cm long, instead of a single large one. Smaller incisions heal faster, require fewer sutures, and are less susceptible to tearing or infection.

An instrument called a laparoscope that has a high-intensity light, and a high-resolution camera is inserted through one of the incisions, which sends real-time images to a monitor. Other incisions provide access points for surgical tools through which the operation is conducted under video surveillance. For most colon or rectal operations, three to five incisions are adequate.

Small tubes, called trocars, are placed at each incision point, through which long, narrow, tubular surgical instruments are passed. Carbon dioxide gas may be used to inflate the abdomen to give the surgeon room to work. A laparoscope is maneuvered through one of the incisions to observe the inside of the abdomen on high-resolution monitors. Surgical implements inserted through incisions are manipulated by the attending surgeon to perform functions such as cutting, clamping, suturing, and cauterizing guided by video images. The operation is typically performed under general anesthesia.

In some instances, due to complications, operations that start out as laparoscopic are converted to traditional open surgery. Wound infections rates in converted cases tend to be higher than in laparoscopic or open surgery cases. Conversion cases also tend to have higher mortality rates. It's therefore important to have a screening method in place to avoid laparoscopic surgery on patients that are at risk of conversion.

Laparoscopic surgery may be used for cancer diagnosis, staging, treatment, and symptom relief. For diagnosis, samples of tissue are cut out from suspected tumors and sent to a laboratory for testing. By examining the inside of the abdomen, a doctor is able to:

- Detect abdominal tumors
- Take samples of abdominal tumors
- Assess the progression of a particular cancer
- Assess the effectiveness of treatment administered

Cancers that can be diagnosed using laparoscopy include:

- Liver cancer
- Pancreatic cancer
- Cancer of the bile duct
- Cancer of the gallbladder
- Ovarian cancer

Laparoscopic surgery is used to excise a wide range of tumors such as gastrointestinal stromal tumors, sarcomas of the abdomen, prostate tumors, and cancerous regions of the colon and rectum. It also can be used to remove an entire bladder or parts of a bladder, lymph nodes, and colon polyps. While there is a clear advantage of laparoscopic surgery over open surgery in the short-term, its routine use as an alternative to long-term curative open surgery for cancer patients is not unequivocally established. Outcome appears to be dependent on the location of the cancer, the complexity of the operation, and the skill of the surgeon. Laparoscopic surgery requires a higher level of skill than open surgery and generally takes longer.

The advantages of laparoscopic surgery over normal surgery are:

- It is less invasive.
- It is less risky.
- It is associated with less pain and quicker recovery times.

For some types of cancer, laparoscopic surgery can be as efficient at removing all the cells of tumors as open surgery, but for others it may be worse. Despite this, laparoscopic surgery appears to be a particularly important option when operating on the elderly or the infirm. In a study of 9400 patients over the age of 70 who had colon cancer surgery, about 60% had open surgery, the other 40% had laparoscopic surgery. Of the patients who had open surgery, 20% were sent to a nursing home compared to 12.5% for those who had laparoscopic surgery. The study also found that patients who had diseases, such as diabetes, heart disease, or high blood pressure, also were more likely to be sent to a nursing home after leaving the hospital.

Laparoscopic surgery is a well-accepted and viable option for the removal of colon cancer. It produces better short-term results and similar long-term cure rates compared to conventional surgery. However, a convincing case for the use of laparoscopic surgery to treat rectal cancer has not been made. Several studies employing highly skilled surgeons have shown better outcomes with open surgery versus laparoscopic surgery for curative rectal cancer procedures. Three separate randomized trials of slightly under 500 patients all showed successful resection rates in the region of 82% for patients undergoing laparoscopic surgery compared to 87% for patients undergoing open surgery. Rectal cancer surgery is more complex than colon cancer surgery, and requires a higher level of skill, which is why many surgeons opt for open surgery when treating it.

Studies on long-term recurrence and survival rates are required for each type of cancer to make convincing arguments for the routine use of laparoscopic surgery over open surgery in the treatment of cancer.

7.2.2 Robotic surgery

Robotic surgery is a variation of standard laparoscopic surgery. The main difference is that instead of operating directly by hand, surgeons use robots to do the work. A computer system translates movements made by a surgeon's hands into movements of robotic surgical instruments. The greater dexterity of a robot arm makes it possible to use smaller tools that make smaller cuts. In theory, with a skilled operator, the finer movements of the robotic arms should deliver greater precision, leading to cleaner removal of malignant tissue, and less damage to nerve tissue and blood vessels than for standard laparoscopic or open surgery. This greater precision should reduce the likelihood of relapses due to missed cancerous tissue. Robotic surgery is considered safe and is associated with lower conversion rates than laparoscopic surgery. Postoperative recovery appears to be similar for robotic and laparoscopic surgery.

Robotic surgery has the potential to broaden the scope for surgeons to perform delicate or complex procedures that would otherwise be tricky or impossible by direct means. For example, it is suited for rectal cancer surgery in the narrow confines of the pelvic region, that contains nerves which control sexual, bladder, and bowel functions, and generally are more challenging. Currently, compared to other procedures, robotic surgery appears to offer great benefits for total mesorectal excision of rectal cancers. As costs decrease, and robots become more sophisticated, the use of robotic surgery may supersede laparoscopic surgery for tricky surgical operations. For this to happen, it must demonstrate better cure rates, and have fewer technical difficulties, as well as an easier learning curve compared to laparoscopic surgery.

Because robotic surgery is relatively new, and the technology is still developing, with limited uptake, a compelling case for its routine use to treat cancers has not been made. A major obstacle is the high cost of the robot and requisite training, which restricts widespread use. Long-term survival data are few and far between, with no clear evidence that robotic surgery provides superior curative outcomes in the treatment of cancer compared to other techniques. Controlled, randomized clinical trials need to be performed to confirm long-term oncological benefits.

As the technology of surgical robots improves, and uptake becomes more widespread, costs should fall and their level of sophistication should improve. Robotic surgery offers the scope for operations to be performed remotely by surgeons sitting in another room, perhaps in a different country.

7.2.3 Mohs surgery

Originally developed by Dr. Frederick Mohs, a U.S. surgeon in the 1930s, Mohs surgery involves the removal of tumors of the skin one layer at a time. A main aim of the technique is to preserve as much nearby healthy tissue as possible, while removing every cancerous cell. Mohs surgery differs from other techniques in that microscopic examination of excised tissues occurs during surgery rather than after, thereby eliminating the need for estimation of the widths and depths of tumors.

A local anesthetic is injected into the skin before surgery. Following removal, each layer of tumor is immediately prepared and examined under a microscope to ascertain if the sample is cancer-free. A minimal amount of normal tissue may be removed with each layer to make sure all tumor cells are excised. Surgery is complete when no more cancer cells are observed, otherwise a further layer of tissue is removed. By this means, Mohs surgery eliminates the guesswork in skin cancer removal. Mohs surgery is routinely used to excise basal cell carcinomas and squamous cell carcinomas, the two most common forms of skin cancer, for which cure rates of 98% have been reported.

7.2.4 Cryosurgery

Cryosurgery has been used to treat skin lesions for approximately 100 years. Despite the name, it is not surgery as it is commonly perceived. Cryosurgery involves the use of very cold temperatures to rapidly freeze and thaw colonies of abnormal cells to destroy them. It can readily be used to manage benign, premalignant, and malignant skin lesions. Following treatment, damaged areas heal naturally as normal cells grow to replace the ones that have been killed. Crusts form on wounds that peels off after a week or two, without the need for any further specialist attention.

To freeze cells a cold substance such as liquid nitrogen is applied as a light spray or via a cotton-tipped applicator. The technique is easy to master and does not require sophisticated equipment or a fully equipped operating theatre. Temperatures of −20°–30°C are effective at treating benign lesions, while temperatures of −40°–50°C are more suitable for treating malignant growths.

7.2.5 Electrosurgery

Electrosurgery involves the use of high-frequency electrical currents to destroy benign and malignant lesions, or to cut tissue. It may be used to kill cancer cells in areas such as skin or the mouth. For surgical purposes, cuts are made by applying intense heat to tissue in the form of sparks generated by an electrode that vaporizes cells. To create a spark, the electrode needs to be slightly away from the tissue. The heat produced during electrosurgery sterilizes the area of the tissue that is being cut, which reduces the risk of infection.

A standard electrical current at a frequency of 60 c/s is unsuitable for electrosurgery because it stimulates body tissue, causes injury, and can cause death by electrocution. High-frequency currents above 100,000 c/s do not stimulate nerves and muscles and therefore, are more suitable for electrosurgery. Special generators are used to take normal current as input and ramp up its frequency to over 200,000 c/s. At this frequency, electrosurgical energy passes through the body with minimal neuromuscular stimulation and no risk of electrocution. Benign growths are often treated with electrosurgery or curettage. Curettage involves the physical scraping of affected tissue. When combined, electrosurgery, and curettage provide an effective means of treating legions and a margin of surrounding tissue.

7.2.6 Laser treatment

The word laser is an acronym for light amplification by stimulated emission of radiation. Normal light is composed of rays of different wavelengths, such as those we see in a rainbow, that have a tendency to spread out. Light used to form a laser is of a single wavelength, which makes it easier to focus as an attenuated beam. A laser is an intense, narrow beam of light of a specific wavelength. Lasers are generated from different sources and with varying degrees of intensities to suit their intended use. Some are powerful enough to cut through materials as tough as steel. The equipment required to generate lasers is bulky and expensive, and requires a certain degree of skill to operate.

In the treatment of cancer, a beam of light is used to perform surgical procedures akin to scraping and cutting with sharp implements. As laser beams can be focused with great accuracy, they are well suited to the task of precision cutting for surgery. A powerful laser destroys cells by vaporizing them. Lasers can cut areas less than the width of a hair, to remove very small cancers with little or no damage to surrounding tissue. Laser therapy is often applied via an endoscope, which is a thin, flexible tube fitted with optical fibers through which light passes. Endoscopes are conveniently directed at target tumors through various openings in the body, such as the mouth, nose, anus, and vagina.

Laser technology permits the use of several non-surgical techniques to remove or shrink tumors with minimal invasiveness. These include:

- Carbon dioxide lasers that remove a thin layer of tissue from the surface of skin along with cancerous and precancerous growths
- Neodymium-doped yttrium aluminum garnet (Nd:YAG) lasers that strike deep into tissue to reach less accessible cancers such as cancers of the airways
- LITT that heat target tissue to temperatures in the region of 60°–110°C, which produces several immediate toxic effects, any of which causes cell death
- Argon lasers that activate drugs fed to cancer cells that kills them, a method known as photodynamic therapy

Good judgment comes from experience,
and a lot of that comes from bad judgment.

Will Rogers

7.2.7 Laser-induced interstitial thermotherapy

The application of heat to tissue causes blood vessel damage and protein structure disruption, leading to coagulation and function loss. These deprive tumors of essential oxygen and nutrients, causing cell death by necrosis. LITT kills tumor cells by applying heat to them via a laser. To improve the positioning and effect of lasers, real-time imaging technology such as MRI, and CT are used. The optimal pointing of lasers at lesions can be guided in three dimensions. The sensitivity of imaging technology to heat permits real-time construction of temperature maps, and the visualizing of induced necrosis of tumors and surrounding areas. LITT technology makes it possible to treat an entire tumor in a single treatment session, and permits the safe destruction of metastases within a safety margin around the lesion. Imaging also helps in the planning and monitoring of follow-up therapy.

A major advantage of MRI-guided LITT is that it can be performed in an outpatient setting, and has a low complication rate. The procedure typically involves a local anesthetic, and takes less than two hours to complete. This reduces hospitalization time and opens up the possibility of offering treatment where other therapeutic options are not possible or have failed.

LITT has been shown to be safe and effective in the treatment of colorectal liver metastases, with superior survival rates to chemotherapy and equivalent to surgery. Experimental data show LITT to be a better option than surgery for patients with colorectal liver metastases, including patients with less than six metastases of a maximum diameter of 5 cm. Any form of therapy that can target metastatic tumors needs to be taken seriously, as they are the main cause of cancer deaths.

Laser therapy can be used alone or in combination with normal surgery, chemotherapy, or radiation therapy. The scope of laser surgery in the treatment and ablation of cancer is becoming broader, and its application growing more efficient as new innovations are brought on-stream. Currently lasers are commonly used to purge small tumors on the surface of tissues, such as skin and the lining of internal organs. Larger tumors are treated by multiple exposures to heat. Lasers also can be used to:

- Relieve bleeding caused by cancer by sealing blood vessels
- Deactivate nerve endings to reduce pain after surgery
- Seal lymph vessels to reduce swelling and limit the spread of tumor cells
- Remove polyps of the colon that may become cancerous
- Shrink tumors causing a blockage of a vital organ such as the esophagus

The advantages of laser treatment over surgery with normal cutting implements include:

- Reduced pain, swelling, and bleeding
- Greater precision
- Reduced risk of infections
- Shorter operating times
- Smaller invasive cuts and scar tissue
- Faster healing times

Cancers being treated by laser surgery include: cancers of the lungs, colon, rectum, cervix, vagina, penis, skin, airway, kidney, brain, liver, and tumors of the head and neck.

Laser surgery is also used for cosmetic purposes such as:

- The removal of tattoos and minor skin blemishes
- The removal of dilated blood vessels from the face
- The removal of unwanted hair

7.2.8 Photodynamic therapy

Photodynamic therapy uses light from a cold laser to activate a drug present at a higher, and therefore more toxic concentration, in tumors than in normal tissue. Unlike chemotherapeutic drugs that kill normal cells, the aim of photodynamic therapy is to selectively target tumor cells.

Photodynamic therapy begins with the administration of a photosensitive drug that is retained at higher levels by cancers cells than normal cells. Following administration, a period of time, which may extend into days, is allowed for the drug to be absorbed and expunged by all cells. Normal cells excrete the drug more efficiently than tumor cells, which consequently end up with higher concentrations of it. A cold laser, such as an argon laser, is used subsequently to irradiate tumor cells at a wavelength that the photosensitive drug can absorb. This energy helps to produce singlet oxygens from oxygen molecules already present in tumor cells. Only light of wavelength up to 800 nm has sufficient energy to promote this photodynamic reaction. The choice of a light source therefore must be harmonious with the absorption properties of the drug. Optimal impact light exposure needs to be carefully timed so that it is applied after normal cells are almost free of the drug, while cancer cells still have sufficient amounts for it to be effective.

Singlet oxygens are very reactive and cause damage to the membranes of crucial organelles inside cells, such as mitochondria and lysosomes. Mitochondria are the powerhouses of cells that produce energy in the form of the molecule adenosine triphosphate (ATP), which is essential for normal cell function. Damage to their outer membranes is bad news for any cell, and sufficient enough to raise alarm bells that trigger apoptosis, the main cause of cell death. Lysosomes contain an array of enzymes capable of breaking down large biological molecules. They degrade material taken up from outside cells and digest unwanted material of the cell. Leaky lysosomes with damaged membranes are also bad news for cells. Necrosis and autophagy-associated cell death are other means by which death may be caused.

As light sources are applied directly to tumors, and normal cells have low amounts of the photosensitive drug relative to tumor cells, very little collateral damage occurs. As photodynamic therapy does not damage DNA, it does not promote the formation of cancer or the development of greater drug resistance by causing mutations as some chemotherapeutic agents do.

Photodynamic therapy has been successfully employed to treat early carcinomas of the oral cavity, pharynx, and larynx while successfully preserving normal functions of speech and swallowing. It also has been used to treat actinic keratosis and basal cell carcinomas. A randomized, placebo-controlled trial comparing photodynamic therapy against surgical excision in 196 patients with superficial basal cell carcinomas showed a 9.3% recurrence rate for photodynamic therapy versus a 0% recurrence rate for surgery after 12 months. Photodynamic therapy is approved for treatment in many countries that include the United States, European Union states, and Canada.

Photodynamic therapy has proven to be a minimally invasive technique that can prolong patient survival who, for various reasons, cannot be operated on, and treatment can make a significant contribution to their quality of life. It also provides an option in cases where mainstream forms of treatment have been exhausted. Although photodynamic therapy provokes an acute inflammatory response apparent as edema at target sites, it does not compromise the immune system, as chemotherapy does, by suppressing the production of white blood cells. The body is not placed under strain to replace vast amounts of cells wiped out by chemotherapy or radiation treatment. Therefore, photodynamic therapy can be used either before or after surgery, chemotherapy, or radiotherapy without severe negative impact. Photodynamic therapy trials with a new drug showed nearly half of a sample size consisting of 413 men had no remaining traces of cancer following treatment.

Lifelong impotence and incontinence are unfortunate side effects of treating prostate cancer with surgery or radiotherapy. Up to 90% of men have difficulty achieving erections sufficient to sustain intercourse, and up to 20% struggle to control their bladders. Importantly, these common side effects were absent two years after photodynamic therapy.

As with MRI-guided LITT, photodynamic therapy can be performed in an outpatient setting, with a low complication rate. With technical innovations and a supply of suitable drugs, there is promise for photodynamic therapy to become integrated into mainstream cancer treatment.

7.2.9 Natural orifice surgery

Natural orifice surgery involves the execution of surgical procedures through natural openings of the body, such as the mouth and rectum. For example, a liver sample may be excised by passing a surgical tool through a small hole made in the stomach having gained access via the mouth. This type of surgery is still at an experimental stage.

7.3 Morcellators

Morcellators are electrical devices used to homogenize tissue or organs into small pieces so that they can be extracted through small incisions during surgery. For example, they are used to remove the uterus of a woman with fibroids by cutting it into pieces and vacuuming out the fragments. While they facilitate the rapid removal of tissue, morcellators spread any cancer cells present to all parts of the abdomen. As there is no way of knowing if cancerous tissues are present before operating, the procedure carries an avoidable risk. Based on current data, the U.S. Food and Drug Administration (FDA) estimates that one in 350 women undergoing hysterectomy or removal of fibroids carry an undetected uterine sarcoma. They have warned against the use of laparoscopic power morcellators for the routine removal of the uterus or fibroids in women. Vaginal hysterectomies lead to similar or better results with a lower complication risk compared to laparoscopic or abdominal hysterectomies.

7.4 Surgery and metastasis

Although surgery substantially improves survival in the early years, there is evidence that it produces a spike in death rates many years later. The processes of taking tissue samples and cutting out whole tumors may themselves promote metastasis. When the survival rate of women with breast cancer who had surgery was compared to those who did not, the group that had surgery showed a surge in death rates eight years later, that did not occur in the group that did not have surgery.

Factors promoted by surgery that may assist cancer cells to bypass natural barriers to metastasis include:

- Cutting open of blood vessels
- Suppression of the immune system
- Growth of blood vessels
- Stimulation of inflammation
- Release of growth factors associated with wound healing

Given the risk, it makes sense for surgeons to take precautions to reduce metastasis during surgical operations.

8 The Cancer Landscape

The genome of a particular individual, the cell type of the tumor, and the combination of accumulated genetic aberrations determine the type of cancer that develops. In any one cell, at any one time, only a subset of the 19,000 genes in the human genome is expressed. A mutation that contributes to the development of cancer in one tissue may not necessarily do so in another. The proteins that the gene codes for it may not be expressed, or they may serve a different function. The characteristics of cancer cells, such as resistance to drugs and virulence, are strongly influenced by cell types, which dictate protein expression profiles, and in turn, influence signaling. How many different types of cancers are there? Why are some more curable than others?

8.1 Control of gene expression

When a cell divides, genes are used as templates to produce its unique set of proteins. During protein synthesis, the base sequences of the genes' DNA are first transcribed into base sequences of RNA, which are then translated into amino acids of proteins. The fine control of this process is critical in determining which proteins are present in a cell at a particular point in time, and in what amounts. The level of each protein is balanced by its rate of synthesis against its rate of degradation. Protein synthesis is regulated, in part, by the initiation and shutting down of transcription, which is the most common means of control. At any given time, each cell type has a unique set of active transcription factors. Some of these increase transcription, while others suppress it. The levels of proteins inside cells can be rapidly decreased by their degradation and by the breakdown of RNA that codes for them.

8.2 Promoters and enhancers

The transcription of genes is initiated by the binding of the enzyme RNA polymerase to a region of DNA known as its promoter. As shown in Figure 8.1, this is located upstream of the start of the gene. Chemical modification of the promoter regions of genes, such as the addition of methyl groups to bases, is used to suppress transcription in part by blocking the binding of RNA polymerase to DNA. This is also shown in Figure 8.1.

There also are enhancer regions of DNA that play an important part in transcription by providing binding sites for regulatory proteins that affect the

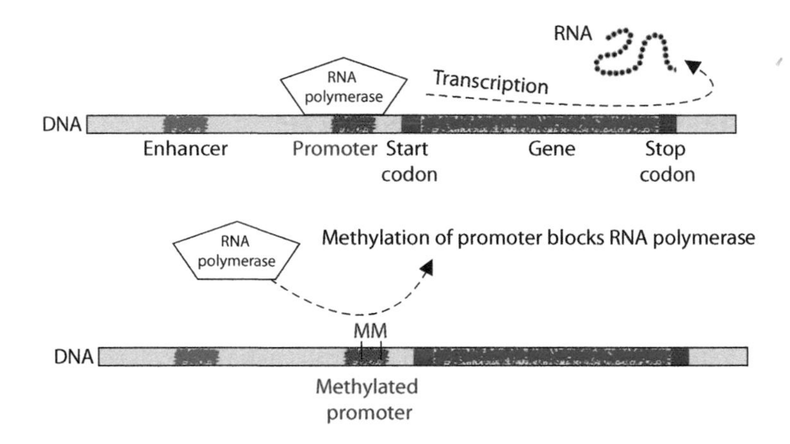

Figure 8.1 Methylation of promoter regulates expression of gene by preventing RNA polymerase from binding to promoter.

activity of RNA polymerase. Enhancer regions may be thousands of base pairs upstream or downstream of a gene being transcribed. The control of the expression of genes, which comes under the umbrella of epigenetic, is primarily influenced by environmental factors.

Transcription factors are a large class of proteins that regulate the transcription of genes. They all have domains that enable them to attach to one or more enhancer or promoter regions of genes. Some transcription factors help to form transcription initiation complexes on binding, and so promote the expression of genes, while other transcription factors block access to promoter regions, thereby suppressing expression of genes. Figure 8.2 shows both promotion and suppression of gene expression by transcription factors. Such modifications of DNA can make critical contributions to the development of cancer, for example, by blocking the synthesis of tumor suppressor proteins.

8.3 Cancer by cell type

All tissues of a fully developed embryo originate from one of three cell layers formed early on in its development:

- Outer layer (ectoderm)
- Middle layer (mesoderm)
- Inner layer (endoderm)

Differentiated cells in an adult that originate from the same layer origin share certain characteristics that influence the type of cancers they form.

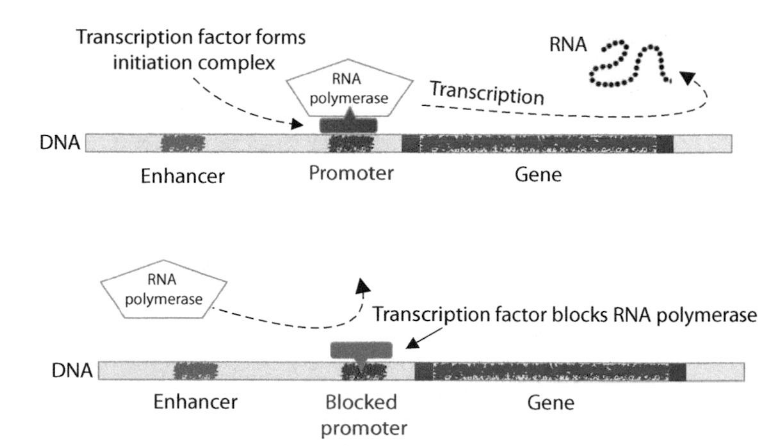

Figure 8.2 Transcription factors regulate expression of genes by forming initiation complexes, or by blocking RNA polymerase from binding to promoter.

Cells that line the esophagus, stomach, intestines, lungs, liver, gallbladder, and pancreas all are derived from the inner cell layer. Connective tissue, bone, muscle, and blood cells develop from the middle layer. The skin and nervous system develop from the outer layer.

There are four basic types of tissue in our bodies:

- Epithelial tissue
- Connective tissue
- Muscle tissue
- Nerve tissue

Epithelial cells are found throughout the body and largely serve to protect underlying tissues that they cover. They are present on the outside of the body in skin, as well as on the inside, where they line blood vessels, cavities, and passages of organs.

8.4 Carcinomas

The four most common cancers occurring worldwide are lung, breast, bowel, and prostate. Though in different locations, all are epithelial cell cancers, which, as a group, account for 80%–90% of all cancers. They are classified as carcinomas. There are two types based on the biological functions of the epithelial cells:

1. Squamous cell carcinomas
2. Adenocarcinomas

8.4.1 Squamous cell carcinomas

One type of epithelial cells forms sheets. These mainly serve to seal the skin or cavity, in order to protect underlying cells. Tumors that arise from epithelial cells of this type are termed squamous cell carcinomas. Cancers of the skin and most of the oral cavity are this type.

8.4.2 Adenocarcinomas

Another type of epithelial cell secretes substances into ducts or cavities that it lines. Often these secreted products are used to protect the epithelial cell layers from the contents of the cavities. For example, some epithelial cells lining the lung and stomach secrete mucus layers that protect them. Tumors that arise from this type of epithelial cell are termed adenocarcinomas.

The epithelia in some organs, such as the lung, uterus, and cervix, have the capacity to give rise to both adenocarcinomas and squamous cell carcinomas. Included among the carcinomas are tumors arising from the lining of the gastrointestinal tract. This includes the mouth, esophagus, stomach, small intestines, and large intestines. Carcinomas also arise from skin cells, mammary glands, lungs, ovaries, the uterus, prostate, gallbladder, urinary bladder, pancreas, and liver.

Why do such a high percentage of tumors arise from epithelial cells? They are especially vulnerable to damage because they are more exposed to radiation and carcinogens. They also undergo more rapid cell turnover than other cell types. This is necessary to negate the accumulation of mutations and to protect against the formation of tumors. Epithelial cells of the colon live for five to seven days before they are induced to die and are sloughed off into the lumen of the intestine. The keratinocytes of our skin die within 20–30 days of being formed, and are shed continuously in the form of small flakes of dead skin.

8.5 World Health Organization (WHO) classification of cancer

The most comprehensive attempt at classifying the different types of tumors has been put together by the WHO. It has published seven volumes that cover over 1000 different tumor types. For example, *WHO Classification of Tumors of the Lung, Pleura, Thymus and Heart* is the seventh volume of the fourth edition. Its coverage includes pathological features, epidemiology, clinical features, microscopy, pathology, genetics, prognosis, diagnostic criteria, and predictive factors. It was produced by 157 authors from 29 countries.

The WHO classifications are updated with the release of each new publication. Tumors are initially grouped by primary site and then by cell type. There are 26 different primary sites:

- Bone, breast, central nervous system
- Digestive system, ear, endocrine organs
- Female reproductive organs
- Hematopoietic lymphoid, heart
- Hypopharynx, larynx and trachea
- Kidney, lung, nasal cavity and paranasal sinuses
- Nasopharynx, odontogenic tumors
- Oral cavity and oropharynx
- Paraganglionic system, penis, pleura
- Prostate, salivary glands, skin

The tissues with the most tumor types are

- Skin: 184
- Hematopoietic and lymphoid: 155
- Soft tissue: 113
- Endocrine organs: 81
- Female reproductive organs: 80

According to WHO classification, there are 12 types of prostate tumors, 12 types of breast tumors, and, surprisingly, 21 types of penis tumors, such a big number for such a little thing.

Most of the primary sites are subdivided into sub-sites. For example, tumors of the digestive system are split into esophagus, stomach, small intestine, and so on. The cell type provides another level of classification. For example, tumors of the esophagus are divided into:

- Squamous cell carcinomas
- Adenocarcinomas
- Neuroendocrine neoplasms
- Lymphoma
- Mesenchymal tumors
- Secondary tumors and melanoma

The impressive classification put together by the WHO has not been embraced by online sources of genetic data including: COSMIC, ICGC, and TCGA. Each uses a slightly different system, which makes it difficult to pool data or carry out comparative analysis. For example, in many cases cancers are classified at the tissue type level, without reference to subtype.

8.6 Breast cancer

It is common to describe breast cancer as consisting of two main types, ductal carcinomas, which develop in the lining of milk ducts, and lobular carcinomas, which develop in lobules. A duct is a tube through which milk is transported from milk-producing glands (lobules) to a nipple. Approximately 90% of all breast cancers are ductal, and 10% are lobular. However, there are other rare types.

While the above method of classification has its benefits, it does not capture the complexity of breast cancers. It also cannot be used as a guide to treatment. The classification of breast cancer using receptors provides an improvement. Chemical messengers such as hormones bind to receptors, following which a number of cellular processes are initiated. Breast cancers are classified according to the presence or absence of three different types of receptors assessed by staining of cells of tumors:

- Estrogen receptors (ER)
- Progesterone receptors (PgR)
- Human epidermal growth factor receptor 2 (HER2)

Depending on the results, cancers are labeled as:

- Triple negative breast cancer
- ER-positive or ER-negative
- PgR-positive or PgR-negative
- HER2-positive or HER2-negative

Cells with none of the three receptors are called basal-like or triple negative breast cancers. Triple negative breast cancers have a poor prognosis in terms of disease-free survival and overall survival. Patients with this condition generally respond to chemotherapy, but relapse quickly. Tumors lacking ER, PgR, and HER2 receptors comprise about 15% of breast cancers.

The majority of breast tumors are ER/PgR-positive and exhibit normal levels of HER2. These have a good prognosis. ER-positive patients account for approximately 65% of breast tumors in women over 50 years of age, while PgR-positive cases account for 60% of breast tumors in general.

We see that breast cancer, which is often thought of as a single disease, is actually several different types. An accurate understanding of the cause of any cancer at the genetic level is necessary for effective treatment.

8.7 Hematopoietic cancers

Hematopoietic cancers are responsible for about 7% of cancer-associated mortality in the United States. They are derived from various cell types used to make blood, and from part of the immune system. Included in this category are red blood cells, lymphocytes, and antibody secreting cells. There are different types of leukemia that include:

- Acute lymphocytic leukemia (ALL)
- Acute myelogenous leukemia (AML)
- Chronic myelogenous leukemia (CML)
- Chronic lymphocytic leukemia (CLL)
- Multiple myeloma (MM)
- Non-Hodgkin's lymphoma (NHL)
- Hodgkin's lymphoma (HL)

It's very likely the formation of hematopoietic tumors requires fewer mutations than solid tumors. First, they do not have to develop their own blood supply. Second, unlike solid tumors, they are not constrained by contact inhibition. Contact inhibition refers to growth cessation following contact with a neighboring cell. This forms one of the barriers a developing solid tumor needs to overcome to attain the characteristics feature of uncontrolled cell division. Cells of rapidly growing tissues in embryos are not strongly contact-inhibited, in contrast to most fully differentiated cells of adults. In cancer cells, contact inhibition may be deregulated, causing differentiated cells to revert to a more primitive developmental state.

Cancers of the blood are generally easier to target with drugs than solid tumors, which may be served by poor blood supply. Survival rates of blood cancers vary substantially according to type, ranging from a five-year survival rate of 26% for patients with AML to 82% for patients with CLL. Some types that had a high mortality rate 30 years ago are now more treatable. For example, the five-year survival rate of ALL has increased from 41% to 70%.

8.8 Chromosome defects and acute myelogenous leukemia

AML affects about 13,000 adults annually in the United States, but treatment has changed little in the past two decades, because we don't understand what causes it. Some forms of cancers are littered with chromosomal defects. For example, over 200 abnormalities of chromosomes and numerous gene rearrangements are associated with AML. Among this chaos, it is difficult to determine which defects are causative and which are symptomatic.

Aberrations at the chromosome level include:

- The production of multiple copies of one or more chromosomes
- The loss one or more chromosomes
- The translocation of part of one chromosome to another
- The fusion of chromosomes

As a consequence of chromosomal translocations, genetic rearrangements may occur that could promote malignant transformation by:

- The placing of the coding sequence of one gene under the control of the promoter of another gene
- The creation of new genes

For a particular cancer, we are not clear on which chromosomal abnormalities contribute to it, which occur as a result of it, or which have no effect whatsoever.

8.9 Chronic myeloid leukemia

In over 95% of patients with CML, part of chromosome 9 and part of chromosome 22 swap places. The new mutant chromosome 22 is known as the Philadelphia chromosome, named after the city in which it was discovered. The fusion of part of chromosome 9, with part of chromosome 22, creates a new gene composed of parts of two different genes: the Abelson kinase (c-Abl) gene of chromosome 9 and the breakpoint cluster region (BCR) gene of chromosome 22. This rearrangement produces a new protein not coded for by our genes, which unfortunately happens to function as a tyrosine kinase that signals cell division, and for which there is no inbuilt regulation in place to shut it down. We should note that the immune system does not destroy cells with this mongrel protein.

In patients with CML, the BCR-Abl tyrosine kinase is active for long periods, causing cells to proliferate at an abnormally high rate. This overproduction leads to levels of immature white blood cells 10–25 times greater than normal. The identification of the major driving factor provided an obvious target for drug development. An inhibitor was duly found. It was named ST1571, but eventually renamed Imatinib (Glivec).

In their search for new drugs, scientists look for molecules similar in structure to the natural substrates of enzymes. The rationale being they are more likely to fit into the pocket of the active site and block out natural substrates. Many drugs work like this. It is easier to find a drug that inhibits the activity of an enzyme than one that activates it.

In a small phase I clinical trial with Imatinib (Glivec) involving 31 patients, all of them experienced complete remission. From a survival period of three to six years, today CML patients taking Imatinib (Glivec) for the rest of their lives survive an average of 30 years after initial diagnosis, thus converting a once rare disease into a relatively common one. There is a lesson here for drug companies.

Some CML patients become resistant to treatment by Imatinib (Glivec) due to a variety of mutations that alter the structure of BCR-Abl tyrosine kinase so that the drug is no longer able to bind to it. To counter this a new drug, Dasatinib (Sprycel), that attaches to a different location, was developed.

This demonstrates a major problem with cancers, they can counter drug therapy over a relatively short period of time by evolving, using the same means by which they become malignant in the first place. Communicable diseases also become resistant to agents such as antibiotics.

8.10 Burkitt's lymphoma

In Burkitt's lymphoma a translocation involving chromosomes 8 and 14 places the microcystin biosynthesis (MYC) gene from chromosome 8 under the control of the powerful immunoglobulin heavy chain gene (IGH) promoter on chromosome 14. This leads to over expression of the MYC protein in lymphoid cells, which promotes cell proliferation.

8.11 Other cancers

Lymphomas include solid tumors formed by the aggregation of B and T lymphocytes, most frequently found in lymph nodes.

Sarcomas are tumors that arise from cells of mesochymal origin. They are cancers of the supporting tissues of the body, such as bone, muscle and blood vessels. Included among these are fibroblasts (the principal cells of connective tissue) and fat cells. Sarcomas make up about 1% of all tumors.

Neuroectodermal tumors develop from various components of the central and peripheral nervous system.

Melanomas are derived from melanocytes, the pigmented cells of the skin and the retina, which arise from a primitive embryonic structure termed the neural crest.

8.12 Brain tumors

Tumors of the brain and central nervous system are relatively rare in the general population. This may be because once fully developed their tissues undergo very little cell division, and are protected by the blood–brain barrier. There are different types of brain tumors that include:

- Gliomas
- Glioblastomas
- Neuroblastomas
- Schwannomas
- Medulloblastomas

Nerve cells of the brain and spinal cord of humans over the age of two show little or no turnover. We are born with the majority of our 100 billion brain cells. A few brain structures add new nerve cells during infancy, but only the hippocampus adds new cells throughout our lifespan. Other animals, such as fish, amphibians, reptiles, and birds display a continuous addition and high turnover of nerve cells in many brain structures throughout life.

Long-term memory is stored as synaptic connections between cells in the cortex of our brain. In the event of cell death in this region, such connections are lost and the memory encoded along with it. It's possible that during evolution our capacity to regenerate new brain cells was given up to permit the preservation of memory. The fact that most of our neurons are as old as we are probably is what allows us to buildup knowledge. A good library and a fast search engine also help.

8.13 Childhood cancers

Cancer is second only to accidents as the leading cause of death in children between ages 1 and 14. According to the American Cancer Society, an estimated 10,380 new cancer cases are expected to occur among this age group in 2016, leading to an estimated 1,250 deaths. Based on the International Classification of Childhood Cancer, cancers common to children and their occurrence rates of all childhood cancers are:

- Leukemia at 30%
- Brain and central nervous system tumors at 26%
- Wilms' tumor (a cancer of the kidney) at 5%
- Non-Hodgkin's lymphoma, including Burkitt's lymphoma, cancers primarily of lymph nodes and bone marrow at 5%
- Hodgkin's lymphoma, cancers primarily of lymph nodes and bone at 3%
- Rhabdomyosarcoma, a cancer of soft tissue at 3%
- Osteosarcoma, a bone cancer at 2%
- Retinoblastoma, a cancer of the eye at 2%
- Ewing's sarcoma at 1%

Although current childhood survival rates vary a great deal according to cancer type and age, they have improved markedly over the past 30 years due to new forms of treatment and improvements over old forms of treatment. Overall, the five-year relative survival rate has increased from 58% in the mid-1970s to 83% for the period 2005–2011. At one end of the scale, five-year survival rates for Hodgkin's lymphomas and retinoblastomas are in the high 90s, while at the other end, survival rates for AML and osteosarcomas are 65% and 69%, respectively.

Recent figures show brain cancers are now ahead of leukemia as the deadliest childhood cancer in the United States. This may be due to the greater accessibility to blood and lymphatic cells compared to brain tissue.

A skull and a blood-brain barrier are special constructs that keep harmful carcinogens away from our delicate and precious brain cells. Unfortunately, they also provide barriers to treatment. Some traditional antitumor drugs are unable to cross the blood–brain barrier, while surgery of brain tissue is dangerous and often impossible. Radiation treatment also poses a problem because of the lasting damage it can inflict. Brain tissue is not able to repair itself the way some other tissues can. We now appreciate pediatric cancers have different genetic causes to adult versions.

9 The Immortal Cell

Embryonic and cancer stem cells have been observed to replicate without limit when cultured in a medium with suitable nourishment and growth factors. In contrast, other cells with the ability to divide stop growing after 40–60 cycles of division, regardless of the contents of their medium. An American scientist, Leonard Hayflick, first discovered this phenomenon, from whose name the term Hayflick's limit was coined. It appears that at some point en route to full transformation into cancer stem cells, normal cells acquire the ability to replicative indefinitely. It also has been observed that normal cells in a culture medium stop growing once they come into contact with each other, a phenomenon referred to as contact inhibition.

9.1 The primordial cell

In theory, a lot of growth can be accomplished within 40–60 cycles of cell division, so is the immortal tag attributed to cancer stem cells justified or even accurate? Let's look at a few basic calculations. Consider how many cycles of cell division it would take to produce two spherical tumors, one with a diameter of 1 cm, and the other a diameter of 5 cm. These would occupy a volume of 0.52 cm^3 and 65 cm^3, respectively. For epithelial tumors, which make up approximately 85% of all human tumors, it has been estimated there are 100 million cells in 1 cm^3 of tissue. Thus, a tumor 1 cm in diameter would be made up of 52 million cells, while a tumor 5 cm in diameter would be made up of 6500 million cells. Assuming there was no attrition, and all members of its progeny can divide, a single cancer stem cell would take 26 cell division cycles to produce the 52 million cells of a 1 cm tumor, and 33 cycles to produce the 6500 million cells of a 5 cm tumor. After 40 cycles of division, the lower end of Hayflick's limit, a tumor 27 cm in diameter can be produced! This would have a volume of 10,300 cm^3 and weigh approximately 10.3 kg. These figures suggest both a 1 cm and a 5 cm tumor could be comfortably produced below the lower end of Hayflick's limit. Why then the need for replicative immortality? Isn't this just showing off? Three possible explanations come to mind:

- A large number of cells die as tumors develop.
- A large number of non-dividing cells are produced as tumors develop.
- During transformation to tumor cells, normal cells regress to an embryonic stem cell phenotype, which comes with built-in replicative immortality as part of the package.

There are two notable factors that could cause the death of tumor cells in large numbers, attack by killer cells of the immune system, and an inadequate blood supply. An early attack by the immune system is very likely to wipe out a newly formed tumor. Whereas, a restriction of blood supply would stall the growth of a tumor. The death of a tumor cell before the tenth cycle of division, when there would be less than 1000 cells, would have a far greater impact on growth rate than the death of cells after 20 cycles of division, when there would be over one million cells.

It is fairly well established that tumors are composed of heterogeneous populations of cells with varying degrees of differentiation. Normal stem cells can divide in two different ways:

- They can form two daughter stem cells both of which can divide.
- They can form one daughter cell that can divide and another cell that is unable to replicate.

Presumably cancer stem cells follow the same pattern of behavior. There is a growing body of evidence that only a fraction of cells in a tumor colony are stem cells. This is the more likely explanation as to why cancer stem cells would need to acquire replicative immortality. The task of tumor growth is restricted to a few select stem cells, the same as for normal tissue types.

We can assume immortality, in the context of tumor development, refers to the ability of some tumor cells to replicate an indefinite number of times. It is a moot point whether or not they produce offspring unable to divide, which are therefore quite mortal. Other than this issue, there are two technical points:

- It has not been demonstrated that cancer cells in a culture are able to divide an infinite number of times as it's not possible to do so.
- Given that each round of cell division produces mutations in the genome of each new cell, the term immortal is a bit of a stretch, as a parent cell, strictly speaking, no longer exists in its original genetic form after dividing.

Given that the backdrop of immortality is infinity itself, it is taking a lot for granted to suggest a cancer stem cell can replicate itself ad infinitum. There is plenty of evidence of what a stem cell can become over an infinite period, assuming 3.5 billion years approximates to infinity. Are we not all the end products of a single primordial rogue cell, with the gift of replicative immortality, unhindered by other cells around telling it what to do? And does this not remind us of a modern-day, rogue cancer stem cell sitting within the confines of a flask somewhere, surrounded by a suitable growth medium? Are cancer stem cells normal cells in which a few biological switches have been flicked that, through sheer chance, cause them to revert back to primordial type, an *infinite* number of years ago?

9.2 Henrietta Lacks's cells

The most famous, and as it turns out, the most controversial cancer stem cells, are those of the Henrietta Lacks (HeLa) cell line. These were originally cultured by George Otto Gey in the 1950s and are still growing today at a steady rate in laboratories all around the world. The original cells were taken from an adenocarcinoma of the cervix of a 31-year-old black woman by the name of Henrietta Lacks, hence the name HeLa. A tumor sample was removed, without her permission or her knowledge, while she was undergoing treatment in the colored ward of the Johns Hopkins Hospital in Baltimore, Maryland, United States in 1951, a sign of the times.

HeLa cells have transformed the biomedical field, and have been used extensively for cancer research and a host of other medical projects. Over 10,000 patents involving HeLa cells have been registered. Today they are sold by the billions without any form of payment to the family or descendants of Mrs. Lacks. Her case has raised legal and ethical issues concerning the rights of an individual to his or her own tissue and genetic data. Neither Mrs. Lacks nor her descendants granted permission for her cells to be harvested, cultured, or sold for profit. In the 1990s, a supreme court in California upheld the legal right of third parties to commercialize cells of Mrs. Lacks without compensation to her family. Johns Hopkins has made it clear they have not patented HeLa cells, or sold them commercially, or benefited by any direct financial means.

9.3 The senescent cell

Once Hayflick's limit is reached, dividing cells undergo a change of state and become senescent, the most prominent feature of which is irreversible growth arrest, brought about by the expression of cell-cycle inhibitor genes and the suppression of cell-cycle promoting genes. The cell cycle consists of a sequence of phases a dividing cell goes through upon receiving signals to divide. Senescent cells have very different characteristics to differentiated cells, which are also unable to divide. Unlike normal differentiated cells, a senescent state is characterized by impaired function. As well as providing a barrier to uncontrolled cell growth, senescence extends beyond tumor suppression into biological processes such as aging, embryonic development, wound healing, and tissue repair. Aging is thought to occur as a result of the gradual accumulation of senescent cells, as we so eloquently demonstrate in our latter years, some more convincingly than others. In theory, cellular senescence should put a limit on the size a developing tumor colony can reach. What's the word on the street regarding it, and how do cancer cells with ninja like skills happen to bypass it?

A cell that has matured to a senescent level differs from its normal counterpart in three ways:

- It cannot be stimulated to enter the cell cycle by growth factors.
- It is resistant to apoptotic cell death.
- It acquires altered differentiated functions.

Recent findings suggest that certain types of DNA damage and excessive growth signaling also trigger senescence. The benefit of this is it provides a barrier that protects against tumorigenic transformation. An aging cell with DNA damage that is unable to divide is not going to take the cancer world by storm, a perfect calm, if anything. There are therefore two types of cellular senescence: replicative senescence triggered on reaching Hayflick's limit and stress induced senescence engendered by DNA damage and excessive signaling. Both are barriers a cancer cell has to negotiate on the road to immortality.

9.4 Cell division

Consider the development of an embryo. The initial fertilized egg divides first into two cells, then four, then eight, and so on until there are roughly 3.72×10^{13} cells in a full-grown adult. After this, further growth ceases and routine maintenance becomes a matter of cell turnover. The daily replacement of billions of cells in the human body is normal as new ones that are fitter for purpose supersede old damaged cells. Cell division is carefully controlled so that there is no oversupply or shortage. Uncontrolled cell growth occurs when cell division is initiated outside of this purview, and when damaged cells do not die as they should.

9.5 Growth factors

To control cell division, cells communicate with each other via growth factors, which as the name implies, signal other cells to divide and grow. Cells also release growth-inhibiting factors. Growth factors are small molecules that are released into the spaces between cells. They then diffuse and bind to receptors on the outer membranes of other cells. Normal cells tend not to have receptors for growth factors they secrete, but some cancer cells do, which provides an independent means of initiating their own cell division. Different cell types may respond differently to the same growth factor. This is because they have an array of receptors on their outer membrane suited to their specific purpose. Moreover, the signaling pathway of each cell type is wired in a manner that suits its precise growth and functional requirements. Many growth factors are versatile, stimulating cell division in a variety of cell types, while others are specific to particular cell types. There are dozens of different growth factors in humans.

9.6 Cell signaling

When a growth factor molecule binds to a receptor, the receptor undergoes a change in conformation that changes its state. An activated receptor in the membrane passes a message to one or more proteins inside the cell that, in turn, pass the message on to one or more other proteins. In this manner, the initial signal may be amplified and branched outwards to strike several different endpoints inside the cell. The matrix inside cells is wired with numerous interconnecting signaling pathways that vary with cell type, the complexity of which is still being unraveled.

Targets at the end of each branch of signaling proteins are effectors, such as transcription factors, that initiate or repress the expression of genes that then go on to execute cellular processes. These may be, but are not limited to, the activation or cessation of cell division, the synthesis of proteins, the initiation of apoptosis, or the ramping up of metabolism to provide energy to sustain cell division. The regulation of transcription is the most common form of gene control.

9.7 The cell cycle

Cell proliferation produces two cells from one. It involves an increase in cell size followed by cell division. Cells with the ability to divide remain in an inactive quiescent state until they receive instructions to proliferate. This inactive state is also referred to as G0. The various messages brought by the various growth factors and growth inhibitors from other cells are weighed up by each cell, from which it reaches a decision whether to:

- Remain in a quiescent state
- Proliferate
- Differentiate
- Undergo apoptosis

The process by which a cell divides, resulting in two daughter cells genetically identical to the parent cell, is called mitosis. This is the final phase of the cell cycle that has four distinct phases that together take about 23 hours to complete. The phases are:

1. G1 for gap 1 (11 hours)
2. S for synthesis (7 hours)
3. G2 for gap 2 (3 hours)
4. M for mitosis (2 hours)

A diagram of the cell cycle is presented in Figure 9.1. Signaling molecules tightly control an ordered progression through the cell cycle. This regulatory

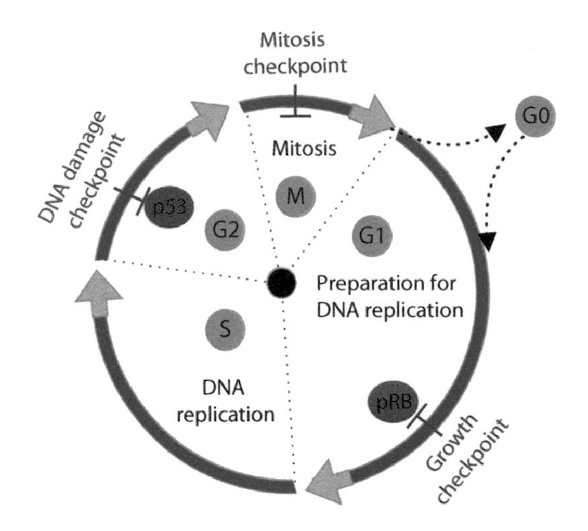

Figure 9.1 Phases and checkpoints of the cell cycle.

system is fundamental to life. Most of its components were in place in primitive organisms such as yeasts, more than one billion years ago. Cell cycle progression is positively and negatively regulated by a family of kinases that turn specific proteins on and off in an ordered sequence. The use of both positive and negative regulators affords finer control of cell cycle initiation and progression. During the gap phases G1 and G2, preparations are put in place for the phases that follow them, and checkpoints are executed to make sure all is well before proceeding. A third checkpoint is carried out in the M phase. As damage to DNA can give rise to any number of diseases, it is imperative that its integrity is maintained. If damage is allowed to pass unchecked from one generation of cells to the next, it's only a matter of time before their accumulation produces cancer and other undesirable diseases. Checkpoints carried out at different points in the cell cycle are shown in Figure 9.1.

9.7.1 The gap 1 phase

During the G1 phase of the cell cycle, copies of organelles are made, accompanied by an increase in cell size as metabolic resources required for the duplication of chromosomes that occurs in the following S phase are built up. Before moving from the G1 to the S phase, a checkpoint is carried out. The cell cycle will stop at the G1 checkpoint, if there aren't enough bases to duplicate DNA, or if it receives signals from surrounding cells telling it not to divide. Cells that have paused will not continue through the cell cycle until they receive signals to continue dividing.

If a G1 checkpoint fails, cell-cycle arrest is triggered, an event mediated by pRB, a tumor suppressor protein coded for by the RB1 gene. Although the mechanism is not fully understood, pRB is believed to limit the synthesis of proteins required for cell cycle progression by forming a complex with transcription factor E2F that binds to the promoters of their genes, thereby reducing their levels of expression and thus their levels of activity, as shown in Figure 9.2. By suppressing transcription targets of E2F, pRB restricts the synthesis of proteins required for cell cycle progression and puts a halt to it. The inactivation of pRB compromises the ability of cells to exit the cell cycle, which is a contributory factor to tumorigenesis.

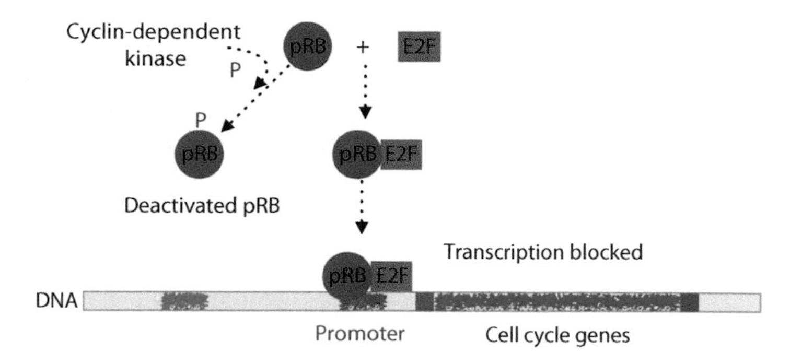

Figure 9.2 Transcription of genes required to drive the cell cycle blocked by pRB-E2F complex. Ability to form complex deactivated by addition of phosphate to pRB.

Protein pRB has been found to be active in quiescent cells, during the G1 phase of the cell cycle, and during checkpoint-mediated cell cycle arrest. The activity of pRB is regulated in part by cyclin-dependent kinases, which deactivate it by adding phosphate groups to it. This is highlighted in Figure 9.2. Cyclins are a family of regulatory proteins that control progression through the cell cycle. They activate cyclin-dependent kinases, by phosphorylating them. They, in turn, phosphorylate other proteins.

An extensive amount of data shows that pRB is functionally compromised in many tumors, either as a result of mutations of the RB1 gene, or via mutations that increase its level of phosphorylation. While pRB regulates progression through the cell cycle in many different cell types, only cells in the retina regularly form cancers when its activity is repressed. It has been shown that gene inactivation of pRB through chromosomal mutations is one of the principal reasons for retinoblastoma development. The inactivation of pRB by viral oncoprotein binding has also been shown in cervical cancer, mesothelioma, and AIDS-related Burkitt's lymphoma.

9.7.2 The synthesis phase

During the S phase of the cell cycle when DNA is copied, the two strands of each chromosome that form a double helix separate. Using each strand as a template, two new complementary chromosome molecules are synthesized, one from each of the original strand. During this process, single bases are joined to a growing DNA chain, one at a time, that match the complimentary base in the template. A key enzyme involved in this process is DNA polymerase, which binds to the promoter site of the gene to initiate synthesis of DNA.

9.7.3 The gap 2 phase

During the G2 phase, preparations are put in place for the mitosis phase that follows it. An important check is the proofreading of DNA to make sure it is accurately replicated and it is not damaged. The cell also synthesizes proteins and other cellular components needed for cell division. Perhaps the most important checkpoint function is the assessment of DNA damage carried out in the G2 phase. If any is found, the cell cycle is put on hold until repairs are carried out. If this is not possible, the cell is destroyed. Apoptosis is the mechanism by which such cells are killed, following checkpoint failure in the G1 or G2 phases. It is checkpoint failure due to irreparable DNA damage that selectively kills dividing cells during chemotherapy and radiation treatment. Quiescent cells and differentiated cells do not die in this manner.

Many types of cancer cells have devised means of bypassing one or more of the checkpoint controls, thus enabling the accumulation of mutations that advance their tumorigenic agenda. Defects in the regulation of checkpoints often result in genomic instability, which accelerates malignant transformation.

9.7.4 The mitosis phase

During the M phase of the cell cycle, the nucleus of a dividing cell splits into two, after which the whole cell forms two daughter cells. A checkpoint is carried out toward the middle of the M phase. To pass it, a cell must have all chromosomes attached to spindles, which are used to pull them apart to opposite ends of the cell prior to splitting. By making sure all chromosomes have spindles attached, the M checkpoint ensures that the two new cells will have the correct number of chromosomes. Checkpoint failure in the M phase of the cell cycle results in cell death by mitotic catastrophe.

9.8 Inside Hayflick's limit

There are two problems concerning the upkeep of the integrity of chromosomes. First, they have free ends. The problem with this is the ends are in danger of being mistaken by DNA repair enzymes as broken and fused together, or with the ends of other chromosomes. Second, DNA polymerase, which catalyzes the copying of chromosomes during cell division, is unable to copy the very ends of chromosomes. The problem this poses is that the gradual shortening of chromosomes with each round of cell division, is accompanied by loss of genetic data. Both of these problems are addressed by telomeres in three different ways.

- First, telomeres are regions of thousands of repeats of the base sequence TTAGGG at the ends of chromosomes. These have no gene-coding or gene expression functions. The loss of pieces of telomeres with each round of cell division therefore is not accompanied by the loss of important data.
- Second, for each pair of chromosomes, one of the strands is a single strand of telomere a few hundred bases in length that folds back onto the double strand to form a T-loop that sequesters it. Distancing the ends of chromosomes in this manner offers protection against them being fused together.
- Third, a complex of six proteins, shelterin, binds to the ends of telomeres and forms a protective cap.

The inability of DNA polymerase to copy the ends of chromosomes results in the loss of 50–200 base pairs of telomeric DNA with each round of cell division. After 60–80 cycles, telomeres shorten from a length of 10–15 thousand bases to a length of five thousand bases or less. This drastic reduction of telomere length impairs the binding of the shelterin cap and threatens the loss of the protection it provides. Immortal stem cells and cancer cells need to repair short telomeres to be able to continue dividing, and to prevent the triggering of senescence. If this does not happen, at some point an alarm is set off. A DNA damage response, facilitated by the protein p53, follows and leads to either cell senescence or apoptosis. If p53 is inactivated by mutation or is absent, this response is mitigated and chromosome shortening continues, shelterin is lost and fusions take place. The fusion of chromosomes is a characteristic feature of many forms of cancer.

The triggering of senescence or apoptosis explains why normal cells in a dish are only able to divide 40–60 times. The initial length of telomeres and their rate of attrition with each round of cell division are parameters that define Hayflick's limit.

9.9 Cell death

Cell death may be caused unnaturally by necrosis or naturally by apoptosis.

9.9.1 Necrosis

Necrosis occurs when a form of trauma, such as physical force, an infection, a poisonous chemical, or a lack of blood supply cause damage to cells. It is a messy process that results in inflammation and stress to the body.

9.9.2 Apoptosis

Apoptosis is an important, pre-programmed biological process that provides a tidy means of eliminating unwanted or damaged cells. For example, as our brains develop, the body creates millions more cells than it needs; the ones that don't form synaptic connections undergo apoptosis to enhance the performance of the remaining cells. The fingers and toes of a developing embryo in the womb are webbed together. Apoptosis is how connecting tissue is removed to create separate fingers and toes.

During apoptosis, proteins, DNA, and other cellular components are destroyed from within by digestive enzymes, such as caspases causing cells to shrink, after which the debris is packaged into small vesicles to be engulfed and cleaned up by cells of the immune system. Fragments of dying cell are not released into the environment and no inflammatory responses are initiated. Thus, unlike necrosis, apoptosis does not cause inflammation or stress.

Apoptosis forms a barrier to tumor development and is a key contributor to the mechanism through which chemotherapy and radiation treatments work. Mutations that render any components of the apoptosis processes ineffective, can lead to survival of cancer cells that would die under normal conditions.

Apoptosis is triggered in cells where DNA damage is beyond repair. Once a cell senses this, it switches into suicide mode, resulting in its death. By removing such cells, apoptosis provides a natural defense mechanism that protects against the transformation of normal cells into abnormal cells, which may eventually develop into cancer cells. Apoptosis may be induced by the binding of death inducing factors to cell surface receptors, called death receptors, or by an internal mechanism. Protein p53, the guardian of the human genome, is the key component of a system that monitors and initiates the repair of damaged DNA, and failing that, the triggering of apoptosis. Not surprisingly, its function is impaired in many forms of cancer, which permits cells with damaged DNA to survive and multiply. Protein p53 is involved in three main functions: growth arrest, DNA repair, and the signaling of apoptosis. It binds to the DNA in the nucleus of cells, which promotes the synthesis of protein p21. This interacts with other proteins resulting in cell-cycle arrest. In cancers cells where p53 is mutated or missing, the synthesis of p21 is not induced, leaving cells without the ability to stop the cell cycle from progressing in the event of DNA damage, which promotes the formation of tumors.

Normal epithelial cells that lose their tethering to the basement membrane activate a form of apoptosis referred to as anoikis. This limits the ability of epithelial cells to migrate to distant sites.

9.10 Immortality

The observation of tumor cells with unlimited capacity to divide suggests the existence of a mechanism to stabilize telomere length. There is an enzyme, telomerase reverse transcriptase (also known as telomerase) that repairs chromosomes by adding back TTAGGG repeats. Telomerase is activated and can maintain telomere length and replicative immortality in embryonic stem cells and cancer stem cells. However, its activity is low or absent in other cells, including the majority of adult stem cells. Although the gene that codes for it is present in all cells, it is not active because it is not expressed. Telomerase is active in 85% of cancers. How do cancer cells manage to turn it on? It appears in a significant number of cases this is achieved by mutations in the promoter region of telomerase that switches on its expression. There are two telomerase promoter mutation hot spots common in cancers that cannot be there by chance, and are therefore culpable.

In some animals such as whales, crocodiles, and lobsters, telomerase is expressed in cells throughout the body. It would be interesting to know their rate of cancer occurrence. Apparently, they do not have clearly defined lifespans and tend to die from diseases and other dangers in the wild.

10 Radiotherapy

If it don't work—hit it.
If it still don't work, use a bigger hammer.

The Birmingham screwdriver

Radiotherapy involves the application of high-energy radiation to targeted areas of the body with the intention of killing tumor cells. As such, it is more suited to tumors that have not spread, so that multiple sites of the body do not have to be treated. In cases where cancers have metastasized, drug treatment provides a better option, as it is able to target dispersed areas of the body. About 40%–50% of cancer patients are given radiotherapy as an adjuvant to surgery in cancer treatment. Also, it is sometimes used in conjunction with chemotherapy to provide a double attack on tumor cells. As previously stated, radiation is used primarily to shrink tumors before surgery or to kill off any remaining tumor cells after surgery. However, it is possible to eliminate some cancers such as early-stage prostate cancer with radiation treatment. The decision on whether to use radiotherapy or not is guided in part by the tumor stage.

10.1 Staging of cancers

The staging of cancers is a means by which practitioners classify tumors according to how advanced they are at the time of diagnosis. It is important to understand that the stage assigned to a cancer is set at the time of diagnosis and does not change with time. If the cancer changes by shrinking or metastasizing, it is still referred to by the stage it was given at the time of diagnosis.

For example, a stage II breast cancer that has spread to bone tissue is referred to as stage II breast cancer with bone metastasis, not stage IV. Statistics on cancer are reported according to stage at diagnosis. There are different staging systems in common use.

Staging information is used to plan treatment, predict outcome, monitor progress, and communicate. Although the underlying genetic abnormalities of each patient's cancer may be different, cancers of the same stage are generally treated alike. The staging of cancers is usually determined by the size of the tumor and the degree to which it has spread:

- To adjacent layers of tissues
- To lymph nodes
- To areas distant from the primary site

Abnormal growths may be described as follows:

- In situ: Abnormal cells are present in the layer of cells in which they developed, and have not invaded nearby tissue.
- Localized: Cancer cells exist only in the organ in which they developed.
- Regional: Cancer cells have spread from the primary site to nearby sites.
- Distant: Cancer cells have spread from the primary site to distant sites.

10.1.1 Stage 0

Stage 0 is used to group cancers in situ, which literally means in place. Removing the entire tumor using surgery is usually the best option for tumors at this stage.

10.1.2 Stage I

Stage I tumors are those that are small, have not grown deeply into nearby tissues, and have not spread to lymph nodes or other parts of the body. They are often referred to as early-stage cancer.

10.1.3 Stage II and III

These two stages are used to classify cases in which tumors are larger and have invaded nearby tissues. They have spread to lymph nodes, but not to other parts of the body. Naturally stage III cancers are bigger and are at a more advanced state than stage II cancers.

10.1.4 Stage IV

This stage covers cancers that have spread to other organs or parts of the body. Stage IV is also described as advanced or metastatic.

10.2 Radiation treatment

Radiation treatment involves the exposure of cancerous and surrounding tissue to high-energy x-rays or gamma rays, which damage the DNA of cells either directly or indirectly. Indirect damage is caused by reactive chemicals formed inside cells by radiation, which produces a variety of lesions in DNA such as:

- Single chromosome strand breaks
- Double chromosome strand breaks
- Alteration of bases
- Crosslinks between bases and formation of dimers between adjacent bases

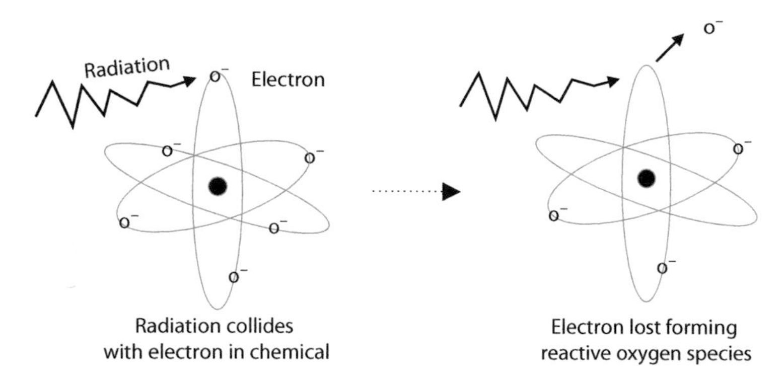

Figure 10.1 Formation of reactive oxygen species by radiation.

The formation of reactive oxygen species by radiation is illustrated in Figure 10.1. These go on to attack DNA causing damage such as those shown in Figure 10.2. Reactive hydroxyl radicals (OH•) produced by radiation cause a high number of single strand breaks to occur.

Cells with DNA damaged beyond a certain point are unable to fulfill their normal function or divide. Dividing cells with severely damaged DNA undergo cell death by apoptosis due to checkpoint failure, while differentiated cells and quiescent stem cells may enter a senescent state. It is important that the dose of radiation administered is high enough to tip cells over the edge so they die. Surviving cells with damaged DNA are good candidates to go on and form tumors as DNA repair is not always carried out accurately. Therefore, radiation therapy needs to find a balance between doses that are high enough to kill all dividing tumor cells, yet low enough to minimize collateral damage. Toward this end, it is important to keep areas of the body exposed to radiation to a minimum, otherwise the high doses of treatment will severely damage or kill the patient.

A machine outside the body generates radiation, or it may come from radioactive material placed in the body near tumor cells. In addition, radioactive

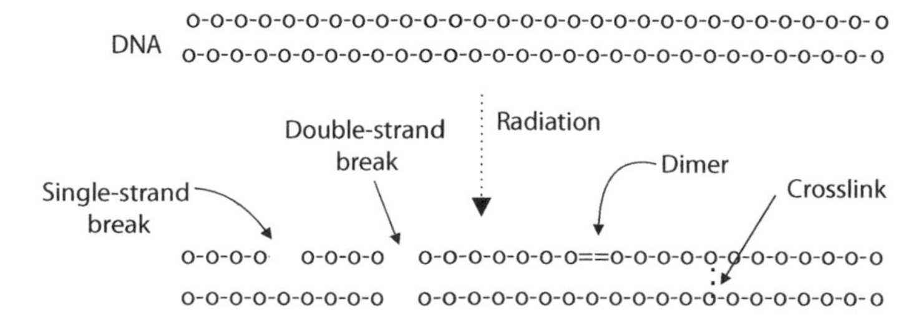

Figure 10.2 Damage caused to DNA by radiation.

agents may be injected into the bloodstream. The location of tumors is pinpointed using a variety of scanning technology: CT, MRI, PET, and ultrasound. Although radiation is targeted at areas with growing tumors, surrounding areas suffer collateral damage.

Radiation therapy causes both early (acute) and delayed (chronic) side effects. Acute side effects occur during treatment, and chronic side effects occur months or even years after treatment. The side effects that develop depend on the area of the body being treated, the daily dosage, the total dosage, the patient's general state of health, and other treatments given at the same time.

Acute radiation side effects are caused by damage to cells in the area being treated. The most common is fatigue, which takes affect about two weeks after treatment begins. This is believed to be caused by the tremendous amount of energy that the body uses to heal the damage done to cells by radiation. Energy levels generally return to normal a few weeks after treatment is completed, but it may take up to a year. Redness and irritation, similar to sunburn, about two to three weeks after the initiation of radiotherapy also is common.

Other acute side effects of radiation therapy are specific to the area being treated. For example, patients receiving radiation therapy to the stomach or abdomen may experience diarrhea, nausea, vomiting, and urinary problems. Medications are available to help prevent or treat nausea and vomiting during treatment.

Patients receiving radiation therapy to the head and neck may experience hair loss, the development of mouth sores, and damage to salivary glands. This results in several adverse secondary side effects, including dryness of the mouth, difficulty swallowing, malnutrition, changes in taste, and increased oral infections. These symptoms are believed to be caused by destruction of the oral tissues. Salivary glands are exquisitely sensitive to radiation. Drug therapy can help protect them from radiation damage if administered during treatment. This works by rapidly reacting with free radicals that are generated by radiation, thereby reducing the damage they inflict on DNA. Tissues such as salivary glands, kidneys, liver, and bone marrow that display a greater uptake of protective drugs than tumor cells benefit from better protection.

Late side effects of radiation therapy may or may not occur. Depending on the area of the body treated, late side effects include:

- Fibrosis (the replacement of normal tissue with scar tissue, leading to restricted movement of the affected area)
- Damage to the bowels, causing diarrhea and bleeding
- Memory loss
- Infertility

A second cancer may be caused by radiation exposure. Second cancers that develop after radiation therapy depend on the part of the body that was treated. For example, girls treated with radiation to the chest for Hodgkin's lymphoma have an increased risk of developing breast cancer later in life. In general, the lifetime risk of a second cancer is highest in people treated for cancer as children or adolescents. It's possible that DNA damage caused by radiation treatment may also create mutations that confer resistance to drug treatment.

Whether or not a patient experiences late side effects depends on other aspects, aside from the dosage of radiation received. Factors such as the chemotherapy drugs administered, genetic risk factors, smoking, and alcohol abuse all increase the risk of late side effects.

Many tumors are served by a poorly constructed blood supply network, causing some of their cells to have relatively low levels of oxygen, a phenomenon known as hypoxia. As radiotherapy works in part by producing reactive oxygen species that go on to damage DNA, hypoxia affords some protection to tumor cells where oxygen tension is low. These areas are likely to be in the center of tumors that are typically less well served by blood vessels. The improvement of blood supply to tumors, thereby raising oxygen levels could help to make radiotherapy more effective. We note that the consumption of large doses of antioxidants as a dietary supplement provides protection to tumor cells by reducing the effectiveness of reactive oxygen species generated by radiation.

For the sake of consistency and ease of comparison, the figures concerning the rates, treatments, and survival rates of cancer presented in this chapter are taken primarily from one source ("Cancer Treatment and Survivorship Statistics" by K.D. Miller et al., published in *CA: A Cancer Journal for Clinicians*, 2016).

10.3 Prostate cancer

It is estimated that there are more than 3.3 million men living with prostate cancer in the United States. The median age at diagnosis is 66 years. If it has not spread beyond the prostate gland, radiotherapy may be used in an effort to cure prostate cancer. This involves giving a high dose of radiation to the prostate gland. The five-year relative survival rate approaches 100% for patients with localized prostate cancer, but drops to 28% for those diagnosed at a distant stage. Over the past few decades, the five-year relative survival rate for all stages of prostate cancer has increased from 83% to 99%. Recent data show 10-year and 15-year relative survival rates of 98% and 95%, respectively. The treatment of early-stage prostate cancer with surgery provides similar survival rates. There is no evidence to support either as the better method.

Many prostate cancer survivors who were treated with surgery or radiation therapy experience incontinence, erectile dysfunction, or bowel complications. A long-term study found that more than 95% of patients who underwent surgery or radiotherapy experienced some sexual dysfunction, while about 50% reported urinary or bowel dysfunction. Post-mortem studies show that 85% of men develop prostate cancer by the age of 85. For reasons that are unclear, incidence rates of prostate cancer are about 60% higher in African Americans than in non-Hispanic whites. Obesity and smoking are also associated with an increased risk of prostate cancer.

10.4 Lung cancer

About 57% of lung cancers are diagnosed at an advanced stage, because early-stage lung cancers are typically asymptomatic. Only 15% of cases are diagnosed at a localized stage, for which the five-year survival rate is approximately 54%. This compares to a five-year survival rate of 27% for regional lung cancer, and 4% for distant stage disease lung cancer. Over the past few decades the survival of lung cancer patients has improved largely because of better use of surgical, chemotherapy, and radiotherapy techniques. More recently the use of targeted drug therapy and immunotherapy has been adopted.

Lung cancer can be broadly divided into three types:

- Small cell lung cancer (83%)
- Non-small cell lung cancer (13%)
- Unclassified (4%)

The five-year survival rates are:

- For small cell lung cancer 7%
- For non-small cell lung cancer 21%

10.4.1 Small cell lung cancer

The name small cell lung cancer is derived from the size of cells as they appear under a microscope. They are minute and mostly filled with a nucleus. It is very rare for someone who has never smoked to have small cell lung cancer.

Surgery is rarely employed as a treatment for small cell lung cancer. This is because it is usually found in both lungs. If it is found in only one lung and has not spread beyond nearby lymph nodes, surgery is an option. This occurs in less than 5% of cases. Surgery for these early-stage cancers is usually supported by chemotherapy, and often by chemotherapy and radiation treatment.

Radiation therapy on its own is typically used for patients where the spread of small cell lung cancer is limited. Where it is extensive, radiation therapy is administered in combination with chemotherapy. Of those treated, temporary remission is experienced:

- With limited spread in 70%–90% of cases
- With extensive spread in 60%–70% of cases

10.4.2 Non-small cell lung cancer

There are three main subtypes of non-small cell lung cancer. Although the cells of each differ in size and shape when examined under a microscope, they are grouped together because their treatment and prognosis often are similar.

About 69% of sufferers of stage I and II non-small cell lung cancers undergo surgery, with approximately 25% also receiving chemotherapy or radiotherapy. Sometimes surgery is used in combination with both chemotherapy and radiotherapy. About 53% of stage III and stage IV patients receive chemotherapy, either with or without radiotherapy.

Advanced non-small cell lung cancer patients are usually treated with chemotherapy, targeted drugs, or a combination of the two. For such cases, chemotherapy in conjunction with radiation treatment may be employed. A large percentage of patients experience remission, but the cancer often returns. Targeted therapy drugs, such as inhibitors of angiogenesis and inhibitors of growth factor receptor, also are used to treat non-small cell lung cancer, as well as immunotherapy drugs that target death cell receptors on T-cells.

Patients with advanced-stage non-small cell lung cancer are treated with:

- Chemotherapy and radiation therapy (35%)
- Chemotherapy alone (20%)
- Radiation therapy alone (17%)

10.5 Breast cancer

For patients with early-stage breast cancer, a mastectomy, which involves removal of all breast tissue, is an option for treatment. Another option is breast-conserving surgery in which only the tumor is removed. In cases where the tumor is large, there are multiple tumors, or there is a reluctance or inability to undergo radiation therapy, a mastectomy is preferable. An increasing number of women demonstrate a preference for a mastectomy over radiation therapy as a precaution against tumor recurrence. An increasing number also choose to have both breasts removed. The proportion of women with non-metastatic breast cancer who elect to undergo a double

11
Driver Mutations

DNA changes that confer growth advantages to tumor cells over normal cells are driver mutations, while those that do not are passenger mutations. Genes that drive the development of cancer are driver genes that may be categorized as oncogenes or tumor suppressor genes (TSGs), according to the functions of the proteins they code for. Among the thousands of different proteins, there are those that:

- Promote cell division
- Sustain cell division
- Keep cell division in check
- Maintain and protect DNA integrity

Typically, oncogenes code for proteins that promote or sustain cell division. TSGs code for proteins that keep cell division in check or maintain DNA integrity. A proto-oncogene is the normal form of an oncogene, when it isn't mutated or overexpressed. Proto-oncogenes become oncogenes when their activities are increased by one or more genetic mutations, or by overexpression. No single oncogene or TSG has been observed to drive the development of all cases of a single form of cancer.

Of the 19,000 genes in the human genome, about 600 have been implicated as cancer driver genes, many not conclusively so. Genes that drive the development of cancer have become apparent primarily from experiments that show they are frequently mutated, overexpressed, or under expressed in different forms of cancer. Although this method readily identifies the common driver genes, it fails to identify rare driver genes, or driver genes that are frequent in rare forms of cancer. To address this, a variety of statistical and computational methods have been used to generate lists of likely candidate driver genes. These employ parameters such as frequency of occurrence, protein function, protein interaction, and conservation of protein structure. The overlap between different published lists of driver genes is typically less than 50%, leading to uncertainty regarding exactly which of the less-obvious genes are driver genes, which are oncogenes, and which are TSGs. The significant differences between published lists of driver genes show that we still have some way to go in the identification of oncogenes and TSGs. The identification TSGs is more difficult because they are complicit by their absence or inactivity.

The Cancer Genome Census, maintained by the Wellcome Trust Sanger Institute in the United Kingdom, is an ongoing effort to manually catalogue genes that have been causally implicated in cancer. Its list currently stands at 616, out of which 250 (40%) are classified as oncogenes or TSGs, and 366 (60%) are not classified at all. Altogether the Cancer Genome Census lists 157 oncogenes and 132 TSGs, which includes 39 genes (6%) that belong to both categories. Of the 616 driver genes, 100 (16%) are associated with germline mutations.

It is easier to inhibit the activity of a protein that is overactive than it is to activate one that is inactive or missing. This is why oncogenes make better drug targets than TSGs. Therefore, correct classification of driver genes offers commercial value.

11.1 Oncogenes

In 1969, Robert J. Huebner and George J. Todaro, working at the U.S. National Cancer Institute, coined the term oncogene to refer to viral information including "that portion responsible for transforming a normal cell into a tumor cell." Despite regular use of the term and recent advances in our understanding of the genetic causes of cancer, there are many different definitions of an oncogene, some of which are not quite accurate. For example, an oncogene is often defined as a mutated gene that turns cell division on, which excludes genes that are over expressed such as the HER2 gene in breast cancer. Some definitions unwittingly include TSGs, while others do not make it clear that an oncogene may be a mutated form of a normal gene.

On top of the layer of genes that make up the human genome sits a layer that controls their expression referred to as the epigenome. This is fluid and changes in response to environmental pressure. Each cell shares the same genome, but each type has a different epigenome suited to its particular function, which may change over a period of time. Cancer is not only caused by the accumulation of gene mutations, it is also driven by the buildup of epigenetic changes.

11.2 An updated definition of an oncogene

Given the lack of clarity concerning the definition of an oncogene, I suggest it be defined as: a gene that codes for a protein that is activated above normal levels by mutation or overexpression, which consequently drives or sustains tumor growth or progression. The term drive refers to the initiation of cell division, while sustain refers to processes that support cell growth and cell division. Progression refers to the advancement of a tumor from one state to the next, such as invasive to malignant. Without the contribution of an oncogene, a tumor would cease to grow or progress to the next state, be it benign, malignant, or metastatic.

This definition permits the inclusion of genes not directly involved in the promotion of cell growth such as those that promote angiogenesis. Without an adequate supply of blood to provide nutrients and oxygen, and to remove waste products, tumors do not grow beyond the size of a pinhead. Growth factors that are expressed above normal levels that subsequently promote angiogenesis in tumors are therefore oncogenes.

11.3 Driver pathways

The formation and progression of tumors are caused by the rewiring of a core number of signaling pathways brought about by the accumulation of genetic modifications that activate oncogenes and inactivate TSGs. In general, signaling pathways are numerous, complex, and poorly understood. A compounding factor is that their composition varies with tissue type. The human signaling pathways data maintained by the U.S. National Cancer Institute has a list of 137 signaling pathways characterized by 9248 interactions. Statistics of pathways constructed and maintained by other organizations vary considerably from this data. The components of the different pathways and how they interact are not well understood.

Enzymes do much of the work involved in the passing on of messages. Inside cells, scattered among the many different pathways, is a class of enzymes known as kinases. These work by catalyzing the transfer of phosphate groups from donor molecules to other proteins, including other kinases. Phosphate groups are small molecules that are ubiquitous to all forms of life. For example, they help to link bases together in DNA, and form part of ATP molecules that serve as the energy currency of life. Once phosphorylated, a kinase typically becomes activated, and when dephosphorylated it typically becomes deactivated. Signals are passed on when kinase A is activated by the addition of a phosphate to it, after which it activates kinase B by catalyzing the addition of a phosphate to it, and so on. At some point soon after activation, kinases A and B are both deactivated by the removal of their phosphates.

A mouse cell is able to transform into a tumor cell by means of an activated Ras protein, and an inactivated p53 protein. We have more safeguards in place, and so need more pathways to be deregulated. We are not certain of the full set for any particular cancer, but a core set that is prevalent would include pathways that control:

- Cell division
- Invasion
- Metastasis

- Angiogenesis
- Apoptosis
- Senescence

We are still learning about the components of pathways, how they are controlled, and how they interconnect. For each pathway, there are multiple candidate driver genes that can cause its deregulation.

11.4 Growth factors

Cell division is controlled by signaling pathways and checkpoints in the cell cycle. It starts with the binding of growth factors to receptors in the outer cell membrane. A cell is only able to respond to a growth factor if it has a receptor to it on the outer membrane. Thus, growth factors demonstrate specificity toward cells types. Epidermal growth factor (EGF) mainly promotes the proliferation of epithelial cells. It is also able to stimulate the growth of some mesenchymal cells and glial cells.

Platelet-derived growth factor (PDGF) primarily initiates the division of mesenchymal cells, such as smooth muscle cells, fibroblasts, and fat cells. It is also able to stimulate the growth of certain types of glial cells in the brain. The PDGF receptor is usually found on the surfaces of mesenchymal cells, but not epithelial cells, in contrast to the EGF receptor, which is usually found on the surface of epithelial cells, but not on mesenchymal cells.

11.5 Pathways that control cell division

The deregulation of pathways that control cell division is common in cancers and very likely features in all of them. Cells release proteins such as growth factors and hormones to communicate with each other, and coordinate their activities. As messages are passed along signaling pathways inside cells, a single protein may activate one or more other proteins, and thereby exert influence on more than one cellular process. Following activation by addition of a phosphate group, signaling proteins are typically switched off by its removal.

Cancer cells in a dish do not need growth factors in their medium to grow, whereas it is necessary for normal cells. There are a number of ways they can influence their own proliferation. They may:

- Acquire the ability to produce and release their own growth factors to which they can respond
- Signal other cells to release growth factors to which they can respond
- Increase the number of receptors in their cell membranes
- Modify receptors in the cell membrane so that they are switched on

- Switch on intermediary signaling proteins in pathways by modifying or overexpressing them

The multiple methods of switching on cell proliferation, combined with the profusion of DNA mutations that randomly occur, makes it very difficult to identify the full set of causative genetic aberrations for any given tumor. Tumor cells displaying the same phenotype may arise out of different genetic events.

11.6 The Ras signaling pathway

The Ras signaling pathway is believed to stimulate:

- Cell proliferation
- Cell cycle progression
- Angiogenesis
- Differentiation
- Cell survival and apoptosis

Ras is a small guanosine triphosphate (GTP)-binding kinase that is a member of a cascade of other similar proteins involved in signal transduction, all the way from receptors in cell membranes to effectors, such as transcription factors, that bind to the promoter and enhancer regions of genes. A diagram of the Ras signaling pathway, which initiates cell division and cell differentiation, is presented in Figure 11.1. In its inactive form, Ras is bound to a guanosine diphosphate (GDP) molecule. It is activated by the replacement

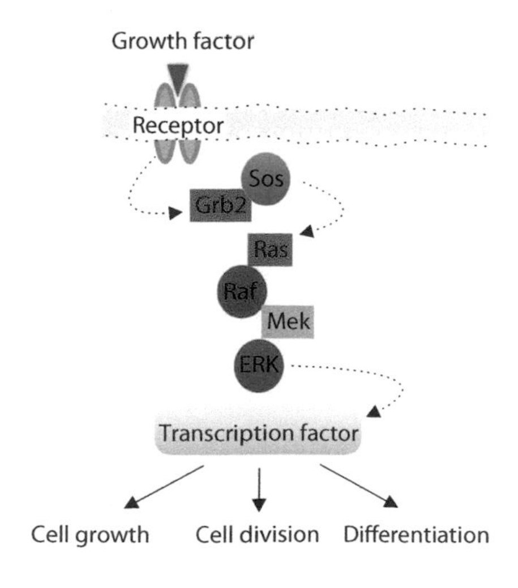

Figure 11.1 The Ras signaling pathway.

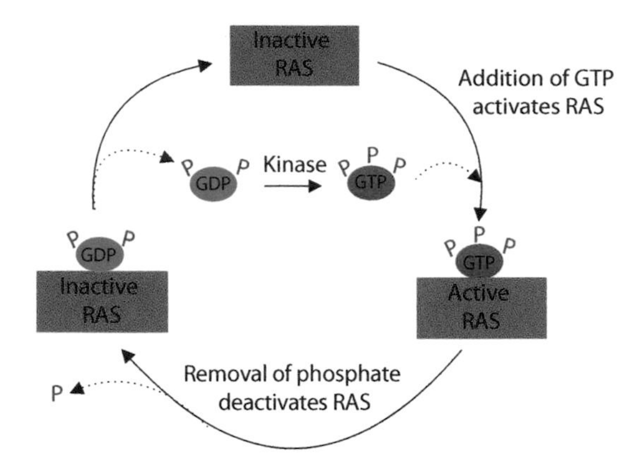

Figure 11.2 Replacement of GDP with GTP activates Ras. Conversion of GTP to GDP deactivates Ras.

of GDP with GTP, and inactivated soon afterward by the conversion of the bound GTP to GDP. This is shown in Figure 11.2. This change is brought about by the removal of a phosphate that is attached to the GTP molecule attached to Ras. When activated, Ras in turn activates a sequence of other signaling proteins. Among these is Raf, which it recruits to the plasma membrane for activation. Signals are passed along the Ras pathway mainly by the addition of phosphate groups to proteins next in line. In normal cells, once switched on, all signaling proteins need to be switched off, otherwise their signaling of processes such as cell division continues indefinitely. Some forms of cancer exploit this weakness by keeping switches in an on state for extended periods of time. Mutations that keep Ras in an on state for extended periods of time are common in certain types of cancers.

Tumors corrupt the Ras signaling pathway at different levels by:

- Increasing production of growth hormones
- Over expression of receptors
- Mutations that activate receptors
- Mutations that leave Ras in an active state
- Mutations that activate Raf

The potential to deregulate the Ras pathway in tumors at so many levels is typical of pathways in general. The observed high frequency of activating mutations of Ras and Raf genes in tumors shows this region of the pathway to be a regulatory hotspot and a potential target for drugs. Further, the observation that Ras and B-Raf genes are rarely mutated in the same tumor suggests Raf is a key signaling protein downstream of Ras. Components of the Ras pathway are targets for the development of cancer drugs.

11.7 The Ras oncogene

Oncogenes of the Ras family (HRas, KRas, and NRas) are frequently activated above normal levels by mutations at codons 12, 13, and 61 of the genes that code for them. KRas mutations are far more frequently observed in cancer, though each oncogene displays preferential coupling to particular cancer types. Mutations corrupt normal signaling by preventing proteins from being switched off following their activation. In some malignant cells a mutation brings about a change in the 3D structure of Ras that prevents the removal of a phosphate from its GTP component to convert it to GDP. This results in it being left permanently in an active state, constitutively signaling a number of downstream biological processes. Studies show that the Ras pathway is able to interact with at least six other, still poorly characterized effectors within the cell. The average life of an activated normal Ras protein is about one minute.

Cancers with the most frequent Ras gene mutations are pancreatic (90%), colorectal cancer (40%), bladder cancer (30%), and non-small cell lung cancer (30%). In contrast, lymphomas, acute lymphoblastic leukemia, hepatocellular carcinoma, osteosarcoma, and prostate cancers all have low occurrences of Ras mutations.

11.8 Tumor suppressor genes

The mode of action of TSGs can be broadly divided into four categories:

- Turn activated oncogenes off
- Repair and maintain DNA integrity
- Trigger apoptosis
- Trigger cell senescence

All of the above suppress the development of tumors. In the absence of adequate DNA repair, mutations accumulate at a faster rate. When this is accompanied by a loss of the ability to trigger apoptosis, a perfect storm is formed that accelerates tumorigenesis. The loss of the capability to trigger senescence is similarly detrimental. Although it is a significant adverse factor, the impairment of DNA maintenance is not necessarily a prerequisite for tumorigenesis.

11.9 An updated definition of a TSG

Given the lack of clarity concerning the definition of a TSG, I suggest it be defined as "a gene that codes for a protein that at normal levels of activity maintains the integrity of DNA, or keeps uncontrolled cell division in check." Aspiring tumors therefore need to deactivate TSGs or suppress their expression. The role of oncogenes and TSGs in the transformation of normal cells into cancer cells is illustrated in Figure 11.3.

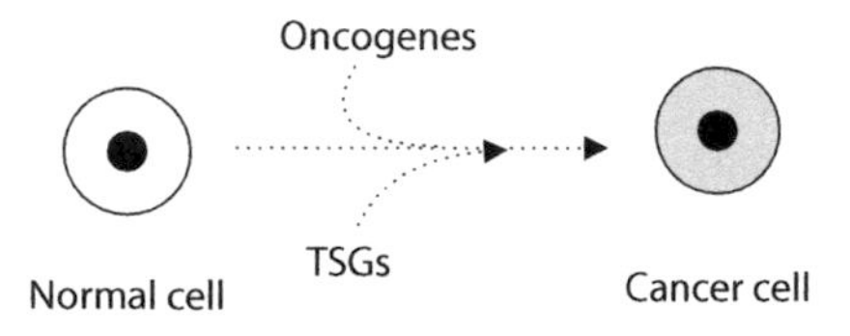

Figure 11.3 The transformation of a normal cell to a tumor cell is caused by the activation of oncogenes and inactivation of TSGs.

11.10 The role of p53

The p53 gene codes for tumor protein p53, which is located in the nucleus of cells throughout the body, and which binds directly to DNA. Loss of one of a pair of p53 genes has been found in virtually every type of cancer. More than 50% of tumors contain mutations of the p53 gene, making it one of the most commonly mutated driver genes. Loss of p53 function increases cell proliferation, decreases apoptosis, and promotes tumor development.

When agents such as toxic chemicals, radiation, or ultraviolet rays damage the DNA of cells, sensor proteins alert p53, which subsequently plays a critical role in deciding cell fate. If damaged DNA is not extensive and can be repaired, it is fixed. Otherwise, apoptosis or senescence is triggered. This prevention of cells with DNA damage from dividing suppresses the development of tumors. The p53 protein works by arresting the cell cycle in the G2 phase.

We don't have a full understanding of the complexities of p53 involvement in mediating cell fate. In normal unstressed cells, p53 attached to mdm2 is maintained at low steady-state levels by its continual breakdown, which restricts its impact on cell fate. In response to DNA damage, p53 is phosphorylated, which causes it to dissociate from mdm2. This reduces the rate at which it is broken down, leading to its accumulation. It is now able to bind to the promoter regions of genes and initiate their synthesis. First, p21 is synthesized, which arrests the cell cycle by inhibiting one or more kinases that drives it forward, as shown in Figure 11.4. Second, if p53 activation continues for a prolonged period of time, the synthesis of other proteins is initiated that induce apoptosis. This ensures cells with damaged DNA, that are not repaired within a given time frame, are killed.

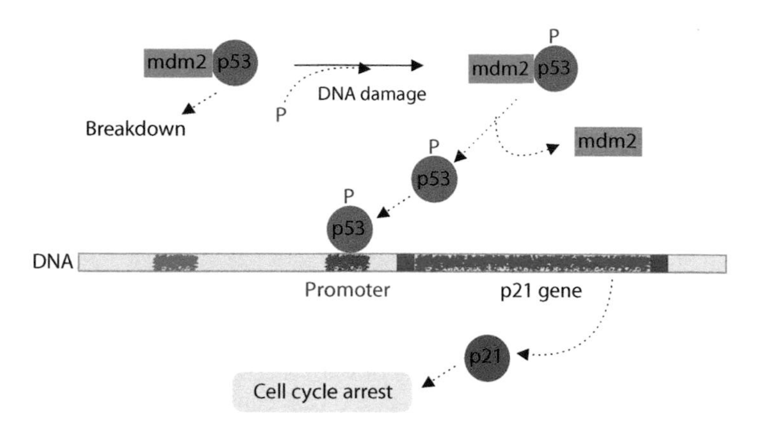

Figure 11.4 Protein p53 is phosphorylated in response to DNA damage causing it to separate from mdm2 and accumulate. This enables it to promote the synthesis of proteins such as p21.

11.11 Li-Fraumeni syndrome

Li-Fraumeni syndrome is a disorder in which there is a predisposition to the development of cancer commonly associated with germline mutations of p53. Individuals who inherit a faulty p53 gene have a 25-fold greater chance of developing a malignant tumor by the age of 50, compared to the general population. Li-Fraumeni syndrome is characterized by the early onset of tumors, multiple tumors within an individual, and multiple affected members of the same family. In contrast to other inherited types of cancer, which are predominantly characterized by site-specific cancers, Li-Fraumeni syndrome presents with a variety of tumor types, the most common of which are leukemia, breast cancer, brain tumors, adrenocortical carcinoma, soft tissue sarcomas, and osteosarcomas.

11.12 Smoking and p53

The association between cigarette smoking and cancer, particularly lung cancer, is well established. It is the main cause of lung cancer, responsible for eight out of 10 lung cancers. Despite this, it remains the most preventable cause of cancer in the world. Lung cancer accounts for more than one in four U.K. cancer deaths, and nearly 20% of all cancers. It has one of the lowest cancer survival rates.

Chemicals in cigarette smoke readily enter the blood stream from where they are transported around the entire body. This is why smoking causes so many different diseases, including at least 14 types of cancer. It has been associated with an increased risk of cancers of the nose, mouth, throat, larynx, esophagus, bladder, cervix, ovary, pancreas, kidney, liver, stomach, colon, and rectum.

Cells lining the alveoli of the lungs cope every day with particulates and pollutants in the air. Cigarette smoke has over 60 carcinogens. At least one of which causes the conversion of the base guanine to thymine in DNA, a process known as transversion. Cigarette smoke can also cause DNA single strand breaks.

In lung cancers, the mutation patterns of genes such as p53 are different between smokers and nonsmokers. The prevalence of guanine to thymine transversions is 30% in lung cancers of smokers compared to 12% in lung cancers of nonsmokers. Mutations of the p53 gene are found in 70% of lung tumors. In response to DNA damage, p53 invokes repair or apoptosis, depending on severity, both of which suppress tumor development. The taking out of this capability by mutations, in conjunction with the accumulation of DNA damage, provides a potent driving force. Research has shown cells damaged by cigarette smoke fail to trigger an apoptotic response. The potency of cigarette carcinogens therefore may result from their ability to damage DNA, while at the same time taking out an important mechanism that protects against tumor development.

11.13 Genetic regulation

There are other important ways in which the activity of driver genes may be regulated, and which may be commandeered by cancer cells to promote and sustain cell proliferation. These include the regulation of protein expression by the:

- Methylation of DNA
- Modulation of the binding of transcription factors to DNA
- Unpacking of DNA around histones
- Upregulation of microRNA synthesis

11.13.1 Methylation of DNA

Methylation of DNA refers to the binding of a chemical group consisting of one carbon atom and three hydrogen atoms (CH3-) to a base. The addition of such methyl groups to DNA is a natural mechanism by which gene expression is controlled. Typically, cysteine bases adjacent to guanine bases are targeted. This results in two methylated cytosine bases sitting diagonally opposite each other on opposing DNA strands. In mammals, global methylation tends to be sparse and evenly distributed, but there are areas 1000 bases or so in length where greater levels of methylation are found. The frequency of methylation near gene promoter sites varies considerably according to cell type. High levels of methylation near the promoter regions of genes correlate with low levels of gene expression. Significant differences in methylation patterns have been observed between normal cells and cancer cells from the same tissue. Methylation is probably used as an injudicious means of silencing TSGs that drive the development of cancer. Although, it has been shown that methylation of the promoter region of telomerase results in increased levels of expression. This may be explained by the prevention of the attachment of transcription factors that suppress the synthesis of the enzyme by methylation of the promoter region.

Methylation comes under the area of epigenetics that relates to the control of gene expression. Epigenetic changes can switch genes on or off to determine which proteins are transcribed in response to environmental factors. Although identical twins may have the same set of genes, differences in their epigenetic makeup means, strictly speaking, they are not genetically identical.

11.13.2 Modulation of the binding of transcription factors to DNA

Transcription factors are protein molecules that bind to DNA and exert positive or negative influences on the expression of one or more genes. A typical transcription factor has multiple functional domains, not only for recognizing and binding to DNA, but also for interacting with other proteins called coactivators. They bind to DNA recognition motifs of the promoter or enhancer regions that are typically six to 10 base pairs long. These regulatory motif sites may be thousands of base pairs upstream or downstream from the gene being transcribed. Regulation of protein synthesis by transcription factors is the most common form of control. Upon binding to compatible sites of DNA, transcription factors may interact with other bound transcription factors and recruit RNA polymerase II. The use of multiple transcription factors to regulate the expression of genes means signals from different sources can be integrated into a single outcome. The action of transcription factors allows for the unique gene expression profile of different cell types, which includes cancer cell types.

11.13.3 Unpacking of DNA around histones

Every cell in the body with a nucleus has about 6 ft. of DNA, which once wrapped around histones, reduces to about 0.09 mm. Histones are proteins that DNA wraps around for the purpose of tighter packing. Chromatin is a combination of DNA and protein, which makes up the contents of the cell nucleus. During protein synthesis, DNA needs to be unpacked, so that enzymes can gain access to the gene template that codes for it. Access also is required for cell division when chromosomes are duplicated. The control of the unpacking of DNA plays an important role in the regulation of gene expression. There are a number of ways that histones can be chemically modified that influence their unpacking and therefore, the expression of genes. These include methylation, acetylation, and phosphorylation.

11.13.4 Upregulation of microRNA synthesis

When proteins are synthesized, DNA is first transcribed into messenger RNA (mRNA), which is then translated into proteins. MicroRNAs are a recently discovered class of small RNA molecules, approximately 22 bases in length, that regulate the translation of mRNA into protein molecules. They are formed by the transcription of genes by RNA polymerases II and III that do not code for proteins. Precursor molecules, approximately 70 bases in length, are first formed by transcription from DNA in the nucleus. They are then transported to the cytoplasm where they are fragmented to microRNA molecules, approximately 22 bases in length.

MicroRNAs work by binding to mRNA, which suppresses protein synthesis, and in some cases, initiates the breakdown of mRNA as shown in Figure 11.5. A single microRNA molecule may be able to bind to as many as 100 different mRNA molecules. In addition, a single mRNA molecule may contain multiple binding sites for many microRNA molecules. This multiplicity of binding combinations results in a complex and interconnecting means of regulating protein synthesis.

MicroRNAs have been found to be involved in the regulation of a wide range of crucial biological processes such as differentiation, cell cycle control, apoptosis, aging, immune response, and viral replication. There is a high degree of structure conservation of in microRNA molecules across different species, indicative of their evolutionary significance as modulators of biological processes. It has been predicted that microRNAs account for up to 5% of the human genome and help to regulate cellular levels of at least 30% of all proteins. A number of tumors have been observed with altered levels of microRNAs. For instance, strong up and down regulation of 16 microRNAs have been observed in primary breast tumors, making them candidate markers for cancer. The targeting of oncogenes and tumor suppressor genes regulated by microRNA represents a novel and potentially powerful therapeutic approach to cancer treatment. The human genome is estimated to code for roughly 2,600 microRNAs molecules.

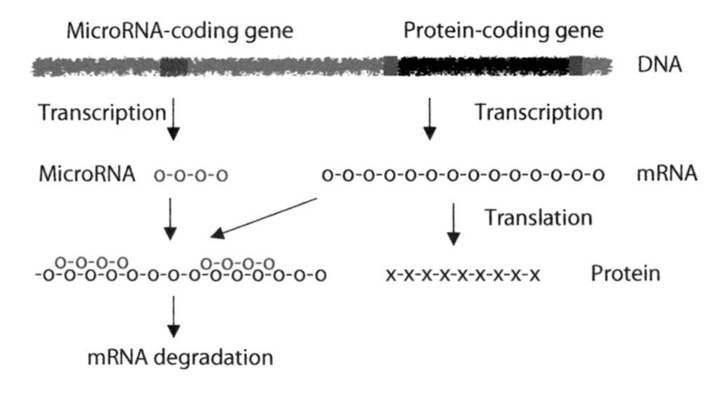

Figure 11.5 Control of protein synthesis by microRNA.

12 Chemotherapy

As cancer is a disease characterized by uncontrolled cell proliferation, in the past drugs produced to combat it mainly focused on inhibiting cell division. Typically, such drugs cause damage to DNA, or hinder the cell cycle at some point, so that it becomes impossible to complete replication. Dividing cells become trapped in an intermediate state and die, enter a senescent state, or regress. Non-specific drugs that target dividing cells are chemotherapeutic agents. They are more likely to be effective at shrinking aggressive tumors, such as acute myelogenous leukemia, than slow-growing ones, such as prostate cancer.

Fully differentiated cells do not divide, and so are spared an abrupt and unceremonious fate as they set about their day-to-day affairs. However, stem cells and progenitor cells are not so fortunate. The cost of killing dividing cells is attrition of precious reserves of stem cells the body requires to replenish old or damaged cells. If you're not on the receiving end, collateral damage is quite acceptable.

Areas that undergo little or no cell division, such as the brain, are less affected by chemotherapeutic drugs. However, tissues that undergo rapid cell division under normal conditions, including the immune system and the digestive system, take a beating. The weakening of a patient's immune system of by such devastation makes them less able to fight off virus infections and to destroy incipient cancer cells.

There are many ways in which dividing cells may be killed, which accounts for the large number of drugs that target different stages of the cell cycle. Currently over 100 drugs have been used as chemotherapeutic agents. Due to the damage they inflict, the use of these needs to be carefully considered, particularly for weaker patients. A recent U.K. study of 23,228 patients with breast cancer and 9,634 patients with non-small cell lung cancer showed 8.4% of patients undergoing treatment for lung cancer, and 2.4% of those being treated for breast cancer, died within a month of commencing therapy.

The Moving Finger writes; and, having writ,
Moves on: nor all thy Piety nor Wit
Shall lure it back to cancel half a Line,
Nor all thy Tears wash out a Word of it.

"The Rubáiyát" by Omar Khayyám

One of the problems with treatments that restrict cell division is that they must be severe enough to take out every malignant stem cell. Not only is this difficult to do without inflicting serious damage, it also is difficult to assess if success has been achieved. This is one of the reasons why relapses occur. It only takes one surviving cell out of billions to compromise the whole effort. It's possible that in a colony of malignant cells, one or more cells may by chance be resistant to the mode of action of a particular drug, and render it ineffective. Alternatively, the treatment itself may promote the creation of new resistant strains of cancer. Common use of the term cancer-free therefore is strange as there is no way of knowing for sure if anyone is actually free of cancer.

Upon treatment, a tumor shrinks as vulnerable cells die. Once this phase is over, resistant cells without competition from other malignant cells, begin to thrive and the cancer remerges. The killing of cancer cells with chemotherapeutic agents comes at a high cost, which includes accelerating the genetic complexity of existing tumors, as well as sowing seeds for new ones. There is madness in the method, and method in the madness.

12.1 Cancer chaos

Cancer is often perceived as a colony of abnormal cells growing rapidly in an uncontrolled manner. This is propaganda spread by those out to give cancer a bad name. The picture is not quite accurate for two reasons. First, tumor growth requires the precise coordination of many complex processes that out of necessity, need to be carried out accurately, and also need to follow the same regulated sequence of events as normal cells. Disruption of the fine level of control beyond a certain point would put an end to cell life and cell division as we know it. Second, tumor cell growth is not as fast as is commonly perceived. There are aggressive tumors that grow quickly, but most take decades to start, and once they do, the average time it takes for them to double in size is more than 100 days. Our most active stem cells divide faster than tumor stem cells.

12.2 Cell replacement

On a daily basis, differentiated cells no longer fit for purpose from normal wear and tear, injury, or disease are killed off to make way for new, fitter cells. It has been estimated that 200 billion new cells are produced every day. The elimination of cells with damaged DNA serves as a major barrier to the development of tumors. Cells that line the digestive system are subjected to harsh conditions such as physical abrasion, stomach acid, and an array of enzymes and chemicals. Not surprisingly, these become damaged and need to be replaced every four days or so. Skin cells last for 21 days. Red blood cells that are squirted around the body through narrow capillary blood vessels at pace last for 120 days before being replaced. These live for a relatively long period of time so their numbers are not as severely depleted by chemotherapy as white blood cells, which take a hit, leading to the risk of infections. In worst-case scenarios, this can lead to life-threatening infections that may need to be treated with antibiotics. Reduction of cell division primarily accounts for the side effects of chemotherapy including sickness, tiredness, and hair loss.

It has been estimated that one in every 10,000–15,000 bone marrow cells is a stem cell. In the blood, the proportion falls to one in 100,000 blood cells. Drugs that exert their effect by killing dividing cells pose the threat of wiping out entire populations of cells capable of dividing. This is likely to be catastrophic to tissues with high cell turnover such as skin, the stomach, and blood, and very detrimental to the immune system. It is imperative that proper limits on drug dosage and durations of treatments are imposed. Fortunately, stem cells are relatively resistant to chemotherapy. In general, they are able to replenish stocks that have been depleted as a result of treatment. Growth factors that specifically promote the proliferation of blood cells also may be administered to boost recovery. In cases where the effects of chemotherapy are more grievous and blood stem cell reserves are severely depleted, bone marrow transplantation is an option. This requires the removal and storage of stem cells from the bone marrow of patients prior to the commencement of chemotherapy.

12.3 Chemotherapy

There are different ways in which cell division may be arrested by chemotherapeutic drugs. They may:

- Modify DNA so that its duplication is impaired
- Restrict the synthesis of bases required for the synthesis of DNA
- Prevent the allocation of replicated DNA to new daughter cells during mitosis

Although a cure is possible for some common tumors using chemotherapy, for most patients it offers a means of slowing down tumor growth or shrinking their size. In the most effective treatments, drugs completely eradicate tumor cells without damaging stem cells, progenitor cells, and normal cells to the extent that they are unable to recover sufficiently to enable patients to return to normal health.

In the absence of a complete cure, achievable beneficial objectives of chemotherapy are:

- Cessation of tumor growth
- Tumor shrinkage

- Prevention or reduction of the invasion of surrounding tissue
- Prevention or reduction of metastasis

As previously mentioned, chemotherapy is often used to shrink tumors before surgery, and to kill off any remaining tumor cells after surgery. Although it has been shown to work well with cancers such as testicular cancer and Hodgkin's lymphoma, its use must be carefully considered. Factors that need to be considered include:

- The stage of the cancer
- The patient's state of health
- The type of cancer
- The patient's quality of life
- The duration of life
- The accurate calculation of dosage

Chemotherapy may produce at least three adverse effects with respect to malignancy:

1. The survival of tumor cells resistant to treatment, that then proliferate in the absence of competition
2. Addition of new mutations to dividing cells making them more aggressive or resistant to further treatment
3. The development of new tumors

12.4 How chemotherapeutic drugs work

Chemotherapeutic drugs are generally non-specific and can cause a lot of collateral damage to other cells in the body. They can be divided into the following six groups according to their mode of action:

1. Alkylating agents
2. Antimetabolites
3. Antifolates
4. Antitumor antibiotics
5. Topoisomerase inhibitors
6. Mitotic inhibitors

12.4.1 Alkylating agents

These are the oldest group of chemotherapeutic drugs. They react with the bases in DNA, as well as a number of other biological molecules, such as RNA and proteins. It is believed their cytotoxic effect is due to the damage they cause to DNA. Alkylation agents add chemical groups to bases of DNA,

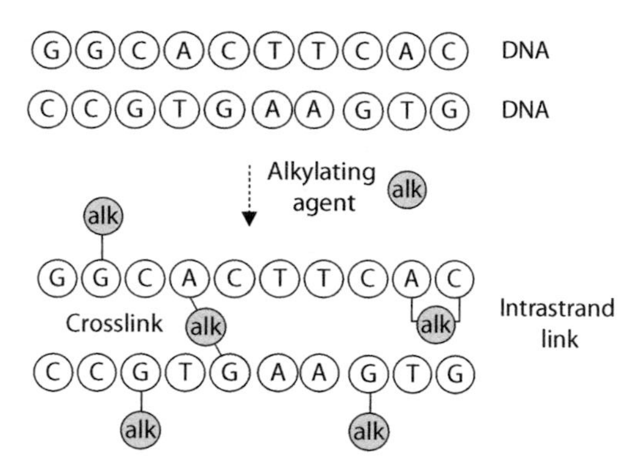

Figure 12.1 Alkylating agents modify bases and form links between bases of DNA.

and also crosslink bases of DNA by forming bridges between those in close proximity. As shown in Figure 12.1, links can be formed between bases of the same strand of DNA and between bases on different strands. DNA repair enzymes are not able to accurately repair damage to DNA inflicted by alkylating agents. Thus, enzymes involved in catalyzing the duplication of DNA during cell division are unable to function normally due to obstructing crosslinks and chemically modified bases. Consequently, severely damaged DNA of dividing cells trigger checkpoint failure, which brings about cell cycle arrest, followed by apoptosis.

Alkylating agents are indiscriminate in that they damage the DNA and other molecules of all cells, whether they are dividing or not. As the collateral damage to the DNA of normal cells exerts a lasting and debilitating effect on their functions, the decision to use alkylating agents must be weighed up carefully against the potential benefits. Where decisions are made to proceed with them, dosages need to be restricted within tolerable ranges of toxicity. In rare cases, long-term damage to bone marrow may lead to acute leukemia. The risk increases with higher doses of alkylating agents.

12.4.2 Antimetabolites

The term antimetabolite is generic in that it refers to any substance that inhibits a normal reaction in the body. In the treatment of cancer, it generally refers to chemicals that inhibit the synthesis of DNA, without which cell division is not possible. Antimetabolites are either analogs of bases, or analogs of molecules required for the synthesis of bases, particularly antifolates. They are used to treat a variety of cancers that include leukemia, breast, ovarian, and gastrointestinal cancers.

DNA polymerases involved in the replication of DNA are one of the primary targets of antimetabolite drugs. There are at least 14 of these enzymes expressed in human cells, three of which are mainly involved in the duplication of chromosomes. The other DNA polymerases carry out functions such as the repair of DNA, the replication of mitochondrial DNA, and the repair of telomeres at the ends of chromosomes. The inhibition of DNA polymerases does not result in the immediate inhibition of cell growth. Following uptake into cells, antimetabolites may suppress the duplication of chromosomes by inhibiting the activity of one or more DNA polymerases during the duplication of chromosomes in the S phase of the cell cycle. They also may be mistaken for bases and become incorporated into the structure of newly synthesized DNA. Due to the absence of the required full complement of chromosomes with the necessary level of DNA integrity, cell division is halted by the checkpoint in the G2 phase of the cell cycle that follows the S phase.

Whereas alkylating agents damage the DNA of all cells throughout treatment, antimetabolites primarily affect cells that are dividing. Further, once a cell has synthesized all the DNA it requires, its DNA suffers very little additional damage from antimetabolites. Thus, beyond a certain dose, cell death plateaus and does not increase in proportion to increased dosage. This makes antimetabolites a less destructive form of chemotherapy than alkylating agents.

12.4.3 Antifolates

Folic acid is a B vitamin prescribed to pregnant women to support a growing fetus. A derivative of folic acid, tetrahydrofolate, is involved in thymine synthesis. As a member of the four bases that make up DNA, any drug that stops its synthesis, puts a halt to DNA replication and cell division.

Antifolates work by inhibiting the activity of enzymes involved in the synthesis of thymine, thereby restricting its supply. Without thymine, dividing cells are unable to synthesize DNA and stop dividing. Unlike other antimetabolites that are base analogs, antifolates are not incorporated into DNA molecules. The fact that they do not compromise DNA integrity makes antifolates a less destructive form of chemotherapy than alkylating agents, or antimetabolites that are analogs of bases. One problem with antifolates is dividing cells can counter their effect by producing more of the enzyme that catalyzes the production of tetrahydrofolate.

12.4.4 Antitumor antibiotics

Antitumor antibiotics are naturally occurring compounds produced by species, such as the soil fungus Streptomyces. Many of these drugs use a variety of means to inhibit cell division. Some are alkylating agents, some insert themselves in between double-stranded DNA molecules, while some promote the formation of reactive chemical agents that damage DNA. Typically, they cause damage to DNA that stops its duplication, and induces cell cycle arrest.

12.4.5 Topoisomerase inhibitors

During the replication of cells, the two strands of DNA of each chromosome pair need to be uncoiled and later recoiled. This involves the breaking and resealing of chemical bonds of single strands of DNA, and the breaking and resealing of bonds of double strands. The enzymes topoisomerase I and topoisomerase II play major roles in these reactions. They also play significant roles in fixing DNA damage caused by exposure to harmful chemicals or UV light.

Topoisomerase I catalyzes the cutting of single strands of DNA in the double helix, while topoisomerase II catalyzes the cutting of both strands. Inhibitors of topoisomerase I and II bind to the enzymes thereby blocking their activity. Depending on the type of inhibitor, and where it acts in the replication process, DNA molecules either are not cut, and therefore not uncoiled, or are not repaired following a cut. The accumulation of these compromises DNA integrity, which triggers cell death by apoptosis.

12.4.6 Mitotic inhibitors

By the time a cell reaches the M phase of the cell cycle, there are two pairs of each chromosome. During mitosis one pair of each is pulled toward one end of the cell, and the other pair towards the other end by microtubules. Most mitotic inhibitors bind to the protein tubulin and inhibit its assembly into microtubules. The disruption of the formation of these brings mitosis to a halt and triggers cell death by mitotic catastrophe.

Microtubules are fibrous, hollow rods that provide structural support and shape to cells. They also function as routes along which organelles can move throughout the cytoplasm of cells. Microtubules are well known to play a key role in the transport of neurotransmitters to synapses, a function required for the transmission of electrical impulses in the nervous system.

Toxicity is a major limitation in the use of mitotic inhibitors. Although they kill tumor cells, they also disrupt the division of normal cells, in particular red blood cells, white blood cells, and blood platelets. Additionally, they also can cause irreversible neuropathy, a condition in which nerves are damaged and sensation, movement or gland function is impaired. Patient resistance to some mitotic inhibitors is commonly observed.

In most cells, the mitotic phase accounts for approximately one of the 24 hours required to complete the cell cycle. Thus, at any one point in time only a small percentage of dividing cells will be susceptible to mitotic inhibition. This makes the task of targeting tumors, especially slow-growing tumors, with such drugs all the more difficult.

12.5 Enhancing treatment

The toxicity and lack of specificity of chemotherapeutic drugs are major concerns in cancer treatment. Another problem is the development of drug resistance. Ideally tumors need to be treated with doses of drugs that are significantly below levels that cause unacceptable side effects. One way to address these problems is to use drug cocktails that administer a combination of drugs with different modes of action. This permits the lowering of doses of individual drugs and reduces the likelihood of the survival of cells that are resistant to a particular drug. A number of different cocktails of chemotherapeutic drugs have been tried with varying degrees of success to treat various cancers.

Most chemotherapeutic drugs work by inducing apoptosis or senescence mediated by the action of p53. The deactivation of this protein is commonly observed in tumors, an event that leads to greater resistant to chemotherapy. Therefore, it may be beneficial to determine if p53 is suppressed in tumors before administering drugs that damage DNA.

The blood vessels of tumors are leaky and poorly constructed, resulting in some cells being supplied with blood at levels that are less than adequate. Unfortunately, this also means that some cells receive lower doses of drugs circulating in the blood stream, which makes them more resistant to chemotherapy. Making blood vessels better at delivering nutrients, oxygen, and drugs to tumors can enhance some forms of cancer treatment. The administration of antiangiogenic drugs that reduce the construction of faulty blood vessels, along with chemotherapeutic drugs, is a means to improve delivery to target cells.

13
Antioxidants

Reactive oxygen species (ROS) are very volatile chemicals that are formed sporadically in the body. They may be generated by external factors such as UV light and x-rays, and internal factors such as the combustion of food. Men in white lab coats with furrowed brows attest that antioxidants protect us against the ravages of free radicals, thereby delaying the onset of chronic conditions such as cancer, atherosclerosis, and aging. The virtues of generous intake of fruits, vegetables, green tea, and a host of off-the-shelf supplements are extolled by manufacturers deeply concerned about the state of our health. Do they have a point, or is it a ruse to separate a fool from his hard-earned money? As we know, they are easily parted.

13.1 Reactive oxygen species

In a similar manner to policemen on patrol, the electrons of atoms and molecules move around in pairs. In the upkeep of law and order, be it civil or chemical, an unpaired agent is an unfortunate wretch that must be repaired as a matter of utmost urgency and concern. Such entities are unpredictable by nature, and attach in desperate fashion to the next compliant respondent they happen upon, unbecoming behavior deserving of more than a critical shake of the head.

All forms of radiation carry energy. When UV rays hit molecules in a cell, their energy separates paired electrons out of their comfort zone, leading to the formation of ROS plus free electron, as shown in Figure 13.1. Damage is done as they react permissively, with unsuspecting molecules such as proteins, cell membranes, and DNA. ROS is a collective term used to refer to chemicals that tend to take electrons from other molecules. They include free radicals and other byproducts of normal chemical reactions that occur in cells. ROS are used by the immune system to kill pathogens that invade our bodies, so neutralizing them is not all good. There needs to be a balance.

If ROS are formed in sufficient numbers to overwhelm the capacity of a cell to deal with them, a condition known as oxidative stress is created. For dividing cells, the typical response is to leave the cell cycle and enter a quiescent phase. With prolonged exposure or high levels of ROS activity, cell death by apoptosis may be triggered. The body produces glutathione, which serves as an antioxidant. The ratio of reduced glutathione to oxidized glutathione within cells is used as a measure of cellular oxidative stress.

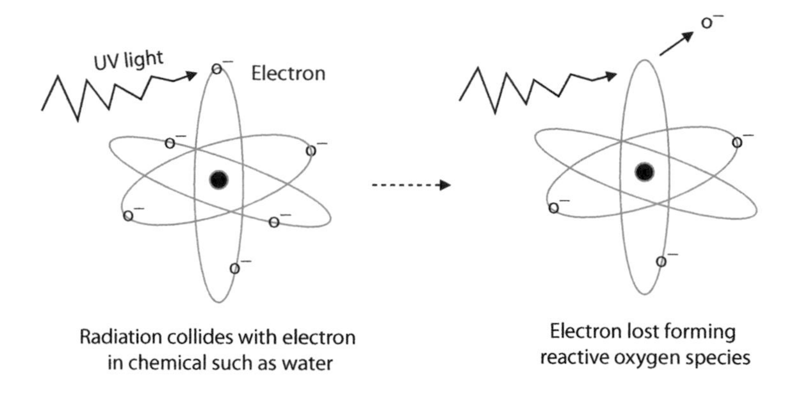

Figure 13.1 Formation of reactive oxygen species by UV light.

13.2 Aging

In the 1950s, U.S. scientist Denham Harman first put forward the suggestion that aging was promoted by free radicals. The accumulation of damage caused by ROS over time also has been suggested as a cause of cardiovascular disease and cancer. Damage to molecules in cell membranes may impact their integrity, while damage to DNA may cause mutations that promote cancer and other diseases.

When DNA is damaged beyond a certain point, cell death is triggered by apoptosis. The peeling of skin, following exposure to sunlight, is an extreme example of the body's response to overwhelming cell damage. Skin cells exposed regularly to high doses of sunlight age faster than normal. This is because they undergo a higher turnover rate, and consequently reach a senescent state earlier. Recall, the telomeres of the chromosomes of cells shorten with each round of division. When they reach a critical point senescence is triggered, and cell division ceases. The accumulation of senescent cells is a major contributor to the characteristics of aged skin.

13.3 Antioxidants

An antioxidant is a chemical that mops up ROS by taking a hit, thereby preventing them from doing damage to other components of cells. There are hundreds of natural compounds in fruits and vegetables that act as antioxidants. Familiar ones are beta-carotene, vitamin C, vitamin E, phenolic compounds, anthocyanins, selenium, and manganese. Vitamins A, C, and E serve as antioxidants, in addition to their other metabolic functions.

13.4 Antioxidants in supplements

The doses of antioxidants in supplements are very different from the levels found in fruits and vegetables. For example, the recommended daily allowance for vitamin E is 22 IU, but an average vitamin E pill contains 18 times that amount. Similarly, a daily diet rich in fruit and vegetables provides around 200 mg of vitamin C, compared to 1000 mg in some supplements. Is it safe to take such high levels of antioxidants on a regular basis?

13.5 Antioxidants and heart disease

Data from some early observational studies suggested antioxidant supplements offered protection against heart disease. However, with the passage of time, this has not been conclusively established. In fact, it could be quite the opposite. Concerning the antioxidant benefits of vitamin A and E, a population study published in 1991 found high plasma levels corresponded with lower rates of cardiovascular disease. Yet, a later, more rigorous placebo-based intervention study of 20,000 adults, showed no impact from high intake of vitamin A and E on cardiovascular disease or death. A number of other rigorous intervention studies using placebo groups equally have been unable to identify any health benefit from taking antioxidant supplements. Alarmingly, cancer, heart disease, and mortality, the very ailments antioxidants are supposed to protect against, actually increased in some studies.

13.6 Antioxidants and cancer

The first large-scale randomized trial to investigate the effects of antioxidant supplements on cancer risk was published in 1993, led by William J. Blot. Supplements that included beta-carotene, selenium, and vitamin E were administered for a period of five years as part of a controlled study involving 29,584 adults aged 40–69. Results showed the taking of antioxidant supplements slightly reduced the risk of developing gastric cancer or esophageal cancer, but this was not conclusively established.

Since 1993, a number of trials lasting between five and 12 years, investigating the effects of antioxidants on the prevention of cancer have been carried out. An examination of nine of these studies showed no benefit from the intake of antioxidants supplements that consisted of beta-carotene, vitamin E, and selenium, either alone or in combination.

13.7 Beta-carotene and lung cancer

A sign that something was amiss with antioxidants came from a large trial led by Gilbert S. Omenn, published in 1996. The study involved a group of 18,314 who were at high risk of developing lung cancer. Included were smokers, former smokers, and workers exposed to asbestos. A placebo-controlled primary prevention trial was conducted in which beta-carotene and vitamin A supplements were administered. Over a period of four years the active-treatment group showed a 28% greater chance of developing lung cancer, and a 46% greater chance of death from lung cancer compared to the placebo group. As a result, the trial stopped early. There were no statistically significant differences in the risks of other types of cancer.

13.8 Vitamin E, selenium, and prostate cancer

A randomized, placebo-controlled trial involving 35,533 men 50 years and over was led by Scott M. Lippman and published in 2009. Researchers examined the influence of selenium and vitamin E on the likelihood of developing prostate cancer and other cancers. The study began in 2001, but was stopped in 2008, approximately five years earlier than planned. Results showed neither selenium nor vitamin E, alone or in combination prevented the development of prostate cancer in the population of relatively healthy men.

Updated findings in 2011, showed 17% more cases of prostate cancer among men who took vitamin E alone than among men who took a placebo. No increase in prostate risk was observed for men who took selenium on its own.

13.9 Avoiding antioxidants

There is no case for the argument that antioxidant supplements prevent cancer. If anything, in sufficiently high doses, they help sustain or, even worse, promote some cancers. Thus, anyone with cancer is well advised to avoid antioxidant supplements. This does not include fruits and vegetables, which are good sources of essential vitamins and minerals, and which are not as overtly rich in antioxidants as off-the-shelf supplements. Fruits and vegetables also may contain other compounds and minerals beneficial to our health. Anyone at a greater risk of lung cancer or prostate cancer should seriously consider desisting from the intake of antioxidant supplements.

13.10 Why don't antioxidants protect against cancer?

If reactive oxygen species damage DNA, it's logical to consider antioxidants that protect against them may well reduce the risk of developing cancer. There are several possible explanations why this is not the case.

First, oxygen tension is lower in the areas of tumors that are not well serviced by blood vessels, such as the center of tumors, which diminishes the potential benefits of antioxidants.

Second, the presence of high levels of antioxidants protects tumor cells, as well as normal cells, by reducing DNA damage below a level that triggers apoptosis.

Third, supplements bolster levels of antioxidants in the blood and cells to unnatural levels that disturb the fine balance between ROS and antioxidants. In so doing, they may suppress the production of ROS that are used by the immune system to kill cells infected with viruses and incipient cancer cells.

Fourth, most of the damage inflicted by UV light is done via direct action on DNA as opposed to the action of ROS. Antioxidants offer no protection against direct DNA damage.

Fifth, high levels of vitamin A hinder the absorption of vitamin D from the intestine by competing for receptor sites. Vitamin D is important for the absorption of calcium and is known to reduce the risk of developing some cancers such as colon cancer. Two studies have reported a doubling of hip fracture rates among women on vitamin A supplements greater than 1.5 mg/day.

Sixth, it's possible that antioxidants assist metastasis. Single migrating cancer cells out in the open in the blood or lymphatic system are exposed to ROS that may be severe enough to kill them under normal circumstances. In the presence of high levels of antioxidants, this effect is diminished, which increases the chances of migrating cancer cells surviving.

13.11 Ultraviolet rays and reactive oxygen species

Excessive exposure to UV rays present in sunlight and artificial light cause skin cancer, one of the most common cancers in the world. Recent cancer registry data by the WHO show incidence is increasing in nearly all light-skinned populations. DNA damage caused by UV radiation is the main cause, which is responsible for over 95% of skin cancers. Increases are observed in all types of skin cancer, including cutaneous melanoma, squamous cell carcinoma, and basal cell carcinoma.

People who tan their skins using artificial UV rays are at a higher risk of developing skin cancer. The younger the age of exposure, the greater the risk. Welders and sheet metal workers may also have a higher risk of melanoma of the eye. Excessive UV exposure also can cause cataracts and suppression of the immune system. The chain of events leading to cell transformation starts with DNA damage caused by UV exposure, followed by the formation of DNA mutations at sites of damage, and then by the accumulation of driver mutations.

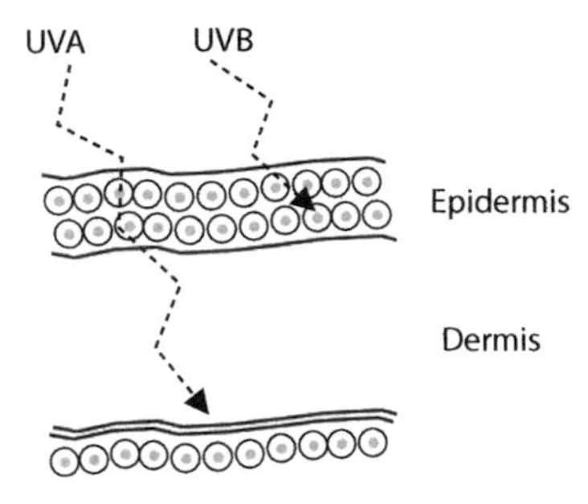

Figure 13.2 Penetration of UVA and UVB rays into skin.

Solar UV radiation can be subdivided into:

- UVA (315–400 nm)
- UVB (280–315 nm)
- UVC (100–280 nm)

The composition of UV light reaching the Earth is:

- For UVA: 90%–95%
- For UVB: 5%–10%
- For UVC: 0%

As shown in Figure 13.2, UVA rays penetrate deep into the epidermis of the skin, compared to UVB rays that only reach as far as the dermis. UVB rays cause sunburn.

13.12 DNA damage

Both UVA and UVB rays cause damage to DNA, by two different means:

1. Directly by the formation of dimers
2. Indirectly by the formation of ROS

13.12.1 Direct DNA damage

Energy from UV rays drives the formation of chemical bonds between adjacent bases in DNA, leading to the production of dimers. These are formed by both UVA and UVB radiation, and are the main means by which UV light inflicts damage to DNA. Dimers are estimated to account for 65% of all UVA-induced mutations, and 85% of all UVB-induced mutations. These cause bulky distortions of the DNA backbone that corrupt the action of DNA polymerase during cell division. For example, if left unrepaired, dimers cause C to T transitions to occur. These are signature patterns of direct damage found almost exclusively in UV-induced skin cancers. The majority of mutations that occur when skin is exposed to sunlight are C to T transitions.

13.12.2 Indirect DNA damage by reactive oxygen species

Energy from UV rays knock electrons out of their normal arrangement, leading to the formation of ROS. These subsequently react with, and cause damage to, DNA. In contrast to C to T transitions produced by direct damage, indirect ROS damage produces G to T and T to G transversions. ROS damage to DNA appears to play a secondary role in UV-induced mutations, which raises questions about the benefit of the widespread use of antioxidants in sunscreen products.

13.13 Xeroderma pigmentosum

Malfunctions that disrupt the repair of CC and TT dimers lead to the rare human autosomal recessive disorder XP. Patients with XP are over one thousand times more likely to get skin cancer compared to the general population. The DNA from squamous cell carcinomas of patients contains a much higher fraction of CC to TT transitions.

13.14 Pigmentary traits

Damage caused by sunlight is dependent on pigmentary traits. For example, dark skin offers greater protection than light skin to sunlight-induced skin cancer. There are wide variations in the toxic effects of UV rays on different members of a population. There also are significant differences in the DNA repair capacity between individuals. Those with the following traits are at increased risk of developing basal cell carcinomas, squamous cell carcinomas, and melanomas:

- Red hair, blond hair, and light brown hair
- Skin that burns but never tans
- Skin that always burns then develops a light tan
- Skin that sometimes burn and always develops a tan
- Skin with freckles on the face, arms or shoulders
- Pale skin
- Blue eyes, gray/green eyes, and brown eyes

13.15 Skin cancer prevention

Taking simple precautions can prevent almost 80% of all skin cancers. Skin protection against UV exposure is an obvious means of reducing DNA damage that promotes skin cancer. Measure to take include:

- Covering skin with clothing and a hat
- Wearing sunglasses with lenses that absorb UV rays
- Avoiding direct sunlight, particularly during the middle of the day
- Using sunscreens with a minimum sun protection factor of 15–20
- Performing regular examinations of your skin

Melanomas occur more frequently on body sites that are exposed to sunlight than on areas that are not exposed. The survival of patients with skin cancer, especially melanoma, is inversely correlated to tumor thickness. Melanomas account for only 2% of skin cancers, but they are the deadliest because they often metastasize. As with all forms of cancers, successful treatment improves substantially with early detection. Optical tools such as dermoscopes are superior to examinations by the naked eye.

14 Vitamin D

Vitamin D is a fascinating vitamin that is quite unique. There are four forms of it in the body, only one of which is active. Unlike any other vitamin, the metabolism of vitamin D follows a circuitous route in the body involving the skin, the liver, and the kidney. What's the story with it, and what does it have to do with cancer?

As far back as 1937, Peller and Stephenson observed skin cancer incidence of men in the U.S. Navy was eight times higher than among men of the same age range in the general population. This is not surprising today given the damage we know sunlight causes to the DNA of exposed skin cells. What was surprising though was that the total death rate among U.S. sailors from cancer of other organs was only 40% of the expected rate. Was there something that offered sailors protection against cancers of organs other than the skin, and if so what was it?

In 1941, Frank Apperley described an association between cancer mortality rates and the location of states in the United States and provinces in Canada. Those who lived further from the equator, such as northern parts of the United States, had higher death rates of colon cancer than those who lived closer. Apperley suggested, "We may be able to reduce our cancer deaths by inducing a partial or complete immunity by exposure of suitable skin areas to sunlight or the proper artificial light rays of intensity and duration insufficient to produce an actual skin cancer." He also noted, "A closer study of the action of solar radiation on the body might well reveal the nature of cancer immunity." This was quite remarkable given the year was 1941, when we had near-zero understanding of the complexities of cancer.

If you're thinking the nature of cancer immunity has something to do with vitamin D, yes, of course, exhibit A. We now know vitamin D is synthesized in the skin under strong, direct sunlight. Of course, there are a host of other factors that cause or protect against cancer, such as exercise, diet, stress, and obesity.

It wasn't until 1980 that a formal connection was published by Cedric and Frank Garland, who suggested vitamin D provides protection against colon cancer. Their study was based on an inspection of the geographic distribution of colon cancer deaths in the United States. They observed mortality rates were higher in places where populations were exposed to the least amounts of natural sunlight.

14.1 About vitamin D

Vitamins are organic compounds that are obtained from food, because we are unable to synthesize them, or because we cannot produce them in sufficient quantities. Without them we become very ill, fail to grow normally, or even die. Of the 13 vitamins essential for normal metabolism, and development, we are only able to synthesize vitamin D and vitamin B3 (niacin).

Four vitamins, A, D, E, and K are fat-soluble, which has two implications. First, the presence of oils and fats in meals is important to enhance their absorption, and second, unlike water-soluble vitamins, they are stored in the body, primarily in fat cells and in the liver. This permits the buildup of reserves of vitamin D during summer months when sunshine is plentiful, and synthesis takes place in exposed skin.

The mechanism of mobilization of reserves of vitamin D, and its significance in satisfying daily requirements is not clearly understood. Fat cells present a problem to obese individuals who require larger doses of vitamin D supplements to achieve serum levels comparable to those of normal weight. This often is not factored into recommended daily allowances of vitamin D.

14.2 Obesity

Obesity is associated with increased risk of a wide range of cancers that includes cancers of the esophagus, pancreas, gallbladder, colon, rectum, thyroid, kidney, breast, and kidney. It is not certain what the cause is, but it could be related to low serum levels of vitamin D. Another suggestion is fat cells produce excess amounts of hormones that affect growth and growth regulators.

14.3 Vitamin D deficiency

Today a number of claims have been made that adequate levels of vitamin D are beneficial for the prevention of a range of health problems, from heart disease and cancer to diabetes. Deficiency of vitamin D is estimated to be present in 30%–50% of the general population. Globally, an estimated one billion have inadequate levels in their blood. Even those considered to be well fed may be deficient in vitamin D. Assuming it offers protection against cancer, a number of questions arise:

- What difference does it make to mortality rates?
- How does it protect?
- What doses of vitamin D provide protection?

14.4 Forms of vitamin D

Vitamin D is a collective term used to refer to any one of four structurally similar substances, only one of which has biological activity. The different forms are

- Ergocalciferol (D2)
- Cholecalciferol (D3)
- 25-hydroxyvitamin D (D4)
- 1,25-dihydroxyvitamin D (calcitriol, D5)

The terms D4 and D5 are not part of standard nomenclature, and are used here for the sake of brevity and simplicity. In food, vitamin D is present in two main forms, D2 from plants, and D3 from animals. These are very similar in structure, and appear to have no biological role in humans, other than to be converted to D4 then D5 in a two-step process. The roles of the various forms of vitamin D are summarized thus:

- D2: Food source from plants
- D3: Food source from animals and also produced in skin
- D4: Inactive form produced from D2 and D3 in the liver for circulation
- D5: Active form produced from D4 in kidneys and some other endothelial cells

The circuitous route of vitamin D is shown in Figure 14.1. In summary, D2 is obtained from plant food sources, while D3 is obtained from animal food sources, and also is synthesized from cholesterol in the skin. Both have no

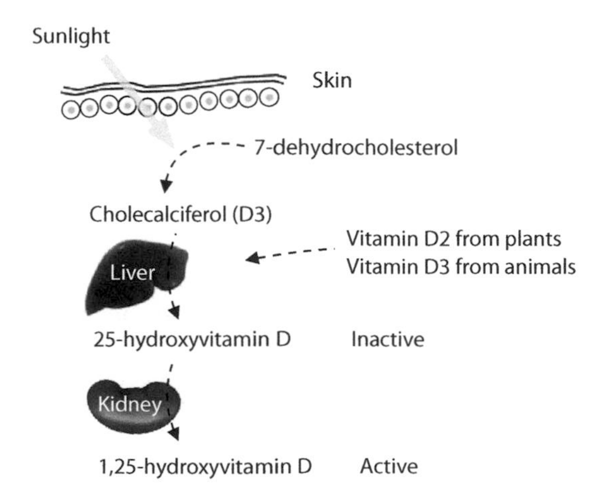

Figure 14.1 Circuitous route of vitamin D.

biological activity and are transported in the blood attached to a carrier protein. D2 and D3 are both converted to D4 by the liver, which is then bound to a carrier protein for circulation in the blood where they serve as a ready reserve. When needed D4, which has no biological activity is converted by the kidney to D5, which is biologically active. It has been shown the body is able to produce D5 from D4 in some endothelial cells outside of the kidney, but the significance of this is not clear.

14.5 Potency of D2 and D3

When taking vitamin D supplements, is it better to take D2 of plant origin or D3 of animal origin? In 2004, Laura et al. showed D3 to be three to 10 times more effective at raising and maintaining D4 serum levels than D2 over a period of 28 days. Their relative potencies were evaluated by administering a single dose of 50,000 international units (IU) of each to 20 healthy male volunteers. This dose is more than five times higher than the recommended daily allowance. Initially, D2 and D3 produced similar rises in serum levels of D4 over the first three days. However, thereafter D4 levels continued to rise for D3 takers, but fell rapidly for D2 takers. D4 levels peaked at 14 days for D3 takers, while it declined to baseline levels for D2 takers. Why the difference in capability between D2 and D3 at raising D4 levels in the blood?

It turns out the carrier protein used to transport D2 and D3 has a much higher affinity for D3, which may explain why D3 was much more effective at raising serum D4 levels. It also may be a contributory factor to the storage of greater amounts of D3 in fat cells. Given its greater potency, it makes sense to favor D3 as a supplement over D2, particularly at high doses.

14.6 Vitamin D and lifestyle

Unlike other nutrients, vitamin D is rare in our diet, so we are unlikely to get sufficient amounts from food alone, regardless of how much we eat. Because of this, certain packaged foods such as cereals and milk are fortified with vitamin D. Fatty fish, such as sardines, wild salmon, herring, and mackerel, are good sources of vitamin D. To a lesser extent, so are egg yolk and cow's liver. Farmed salmon, a fish that is widely consumed in the United States, contains about 25% of the vitamin D3 content of wild Alaskan salmon. That's about it as far as obtaining significant amounts of vitamin D naturally from our diets.

Some standard lifestyle advice handed out by health gurus can lead to a reduction of vitamin D levels in the body:

- A reduction of the intake of fatty foods
- The use of drugs to lower blood cholesterol

- The recommendation of doses of vitamin D supplements that are not high enough
- The avoidance of sunlight
- The use of sunscreen

The synthesis of D3 in skin is dependent upon adequate levels of cholesterol in the blood, as well as sufficient sunlight. As foods low in cholesterol are also low in vitamin D, low-cholesterol diets are inherently deficient in vitamin D.

14.7 Synthesis of vitamin D

Vitamin D3 is synthesized in skin in the presence of UVB rays as follows:

- Cholesterol circulating in the blood is absorbed by cells of the skin.
- Cholesterol is converted to 7-dehydrocholesterol (7-DHC) by the action of an enzyme.
- The energy of UVB rays converts 7-DHC to vitamin D3.

Melanin pigment in dark skin blocks UVB and therefore limits production of D3, as do clothing and sunscreen. It is a combination of the above factors and winter months, plus the low levels of vitamin D in most foods, that make its deficiency so prevalent.

14.8 Vitamin D and cancer risk

A large and growing number of epidemiological and observational studies show that adequate levels of vitamin D intake or synthesis reduce the risk of developing various cancers. In other words, a high level of serum vitamin D somehow provides protection against cancer. Various researchers have highlighted an association between UV light and cancer mortality rates for cancers of the bladder, breast, colon, kidney, lung, esophagus, ovary, pancreas, prostate, rectum, stomach, and uterus. In 2005, Esther M. John led an investigation into the relationship between sun exposure and prostate cancer. The study involved 450 men with advanced prostate cancer and 450 unaffected men. The researchers found those with a high level of sun exposure exhibited a 50% lower prostate cancer risk than men with low sun exposure. Research teams led by Edward Giovannucci and Michael F. Holick found that low serum vitamin D3 levels were associated with high incidences or increased death rates of cancers of the colon, prostate, and breast.

In 2005, Gorham et al. quantified the risk of colorectal cancer to it being reduced by 50% at serum D4 concentrations of 80 nmol per liter. A Canadian study of 1200 women found those who regularly took a vitamin D supplement cut their cancer risk by 60%. Low intakes of calcium and vitamin D have been associated with an increased risk of recurrence of precancerous polyps of the colon. Therefore, optimal vitamin D and calcium status may be important for the prevention of colon cancer.

Although significant and highly indicative, epidemiological and observational studies do not conclusively establish the benefits of vitamin D in the manner of formal clinical drug trials. As previously mentioned, other factors such as diet, exercise, obesity, and stress could be causative.

The first rigorous clinical study was carried out by Wactawski-Wende et al. in 2006. A randomized, double blind, placebo-controlled trial involving 36,282 postmenopausal women was carried out to determine the benefits of vitamin D and calcium supplementation on the incidence of colorectal cancer. Doses of 200 IU vitamin D and 500 mg calcium were administered. It was found that daily supplementation over a period of seven years had no effect on the incidence of colorectal cancer. Was the dose of vitamin D in this particular trial high enough? Nevertheless, a highly significant inverse relation between baseline D4 serum levels and incident cancer risk was noted. Recommended daily allowance of vitamin D is now around 1000 IU compared to 200 IU used in the study.

In contrast to the findings of Wactawski-Wende et al., studies by Lappe et al. published in 2007 did find a reduction in cancer risk. Doses of 1100 IU vitamin D and 1400–1500 mg calcium were administered, much higher than amounts used by the Wactawski-Wende team. An intervention study to determine the efficacy of calcium alone, and calcium plus vitamin D, in reducing cancer risk of all types was carried out. A population-based, double blind, randomized, placebo-controlled trial was conducted on 1179 community-dwelling women randomly selected from a healthy population of postmenopausal women over 55 years. It was found improving calcium and vitamin D nutritional status substantially reduced cancer risk in postmenopausal women.

Fifty women developed non-skin cancer during the course of the study, 13 of which occurred during the first year. In comparison to the placebo group, the relative risk of developing cancer was 40% for the calcium and vitamin D group and 53% for the calcium only group. If it is assumed that cancers diagnosed early in the study were already present and unrecognized, the relative risk for the calcium and vitamin D group drops to 23% and that of the calcium only group rises to 59%. In addition, Lappe et al. also concluded serum D4 concentrations were a significant, independent indicator of cancer risk.

14.9 How does vitamin D offer protection?

It has been suggested in various publications that vitamin D plays a role in:

- Decreasing cell proliferation
- Inducing differentiation of epithelial cells

- Preventing metastases
- Preventing angiogenesis
- Promoting apoptosis

This is an impressive array of effects, all of which hinder the development of tumors. Nearly all cells in the body have receptors to vitamin D, which is indicative of its widespread influence on cellular processes.

In contrast to other vitamins that are involved in the catalysis of metabolic reactions, vitamin D functions as a hormone. Its structure is based on cholesterol, the same as other hormones, such as estrogen and testosterone. Its active form is produced at one location, and its effects are exerted at one or more other sites, the same as for other hormones.

One major role of vitamin D is the regulation of calcium and phosphorus concentrations in serum. It assists in the absorption of these minerals in the small intestine. When serum levels are low, it is also involved in the mobilization of calcium reserves from bones into serum for widespread use.

Calcium is most noted for its role in the building of healthy bones, muscle contraction, and the transmission of messages in the nervous system. It also has been noted to be involved in:

- Cell adhesion
- Cell division
- The catalysis of reactions
- Blood clotting
- The relaxation and constriction of blood vessels
- The secretion of hormones such as insulin

Vitamin D, which helps to regulate levels of calcium in blood, is indirectly implicated in the above functions. Of particular relevance to cancer are cell division and cell adhesion, which help to organize various tissue types and organs in our bodies.

14.10 Cell adhesion, invasion, and metastases

There are molecules that perform cell adhesion in a highly selective manner so that each cell only sticks to other cells it is supposed to. Cadherin proteins form a particular family of adhesion molecules that only glue cells together in the presence of calcium. As cell separation must take place to create space for tumor cells to grow, it has been suggested that vitamin D may hinder the development of cancer by maintaining optimum levels of calcium, and thereby maintaining cell adhesion. If cells cannot separate, metastases cannot take place, and cell invasion becomes difficult. Cell adhesion is therefore important in keeping metastases in check. Several reports demonstrate an antimetastatic effect of vitamin D3 compounds. The underlying mechanism is not clear.

14.11 Cell division and differentiation

Cell lines are tumor cells that are cultured in laboratories under conditions that favor their growth. It has been shown that the active form of vitamin D, which I refer to as D5 (calcitriol or 1,25-dihydroxyvitamin D) inhibits the proliferation of human melanoma cells in vitro. Growth inhibition also has been observed in several other human cancer cell lines such as osteosarcoma, breast cancer, bladder cancer, and prostate cancer.

D5 and D3 analogs have been shown to block cell cycle progression of tumor cell lines, such as prostate, breast, colon, osteosarcoma, malignant myeloma, and squamous cell carcinoma. The inhibition of cell growth may be due in part to the stimulation of cell differentiation. For example, treatment of human prostate cancer cells with D3 analogs led to a more differentiated phenotype that expressed prostate specific antigen and a prostate specific enzyme. Also, osteosarcoma cells produce bone matrix proteins when treated with D5 and D3 analogs.

14.12 Apoptosis

D5 and its analogs have been reported to induce apoptosis for a variety of tumor cells that includes cells of the bladder, bone, breast, colon, lung, prostate, plus melanoma cells, myeloma cells, and leukemia cells. The molecular mechanism of the influence of D5 on apoptosis is not clear.

14.13 Recommended doses of vitamin D

A clinical laboratory test is available to identify vitamin D deficiency. The level of 25-hydroxyvitamin D, which I refer to as D4, in serum is used as a marker, where vitamin D status is broadly categorized as follows:

- Deficient < 30 nmol/l (12 ng/ml)
- Sufficient > 50 nmol/l (20 ng/ml)
- Optimum 80 nmol/l (32 ng/ml)
- Adverse effects > 125 nmol/l (50 ng/ml)

One nm/ml equates to 2.5 nmol/l. The figures above do not factor in the amount of vitamin D that is stored in body tissues.

Vitamin D levels less than 30 nmol/liter are associated with rickets in infants and children, and osteomalacia in adults. It has been estimated a serum level of 50 nmol/l can be achieved in most healthy adults between the ages of 19–70 by the intake of 600 IU of vitamin D per day. Those at risk of low serum levels are advised to take vitamin D supplements or spend extra time in the sun.

As vitamin D is fat-soluble, it's best to take supplements of it with the meal containing the most fat, which is most likely your largest meal of the day. This is important as it could make a big difference to amounts absorbed.

One microgram of vitamin D equates to the biological activity of 40 IU. The recommended daily allowance of vitamin D varies with the recommending body, and ranges between 400 and 5000 international unit (IU) per day! Although there is ridiculous variation, the trend is on an upward curve. The dangers of overdosing on vitamins, which can be very serious, are less with vitamin D than with other vitamins and minerals. High levels of vitamin D should thus be taken on its own, not as part of a general supplement.

In cases of severe deficiency, it is quite feasible to boost vitamin D intake for a short period of time. It has been suggested an upper limit be increased from 2000 to 4000 IU per day. The impact of long-term exposure to high levels of vitamin D is unknown, so high doses should not be taken for long periods of time. Once serum levels of D4 have normalized, normal levels of intake should be administered.

Currently 600 IU per day is the most recommended dose for anyone between the ages of one and 70. Seniors over 70 require more vitamin D as their skin productivity is less. Figures in the region of 800 IU per day have been suggested for the elderly. Recent analysis supports the notion that the recommended daily allowance of 600 IU for vitamin D is too low. For example, in 2015 Veugelers et al. stated their findings following the examination of data from 108 published estimates of vitamin D supplementation and status from a total of 11,693 participants. From their analysis, they recommended daily intake as follows to achieve serum D4 levels of 50 nmol/l or more:

- Normal weight, 2000 IU
- Overweight, 2800 IU
- Obese, 6200 IU

For obese individuals, the figure is 10 times more than the current recommended daily allowance! A daily allowance of 1000 IU for adults, who are not obese, seems about right.

Generally, up to 95% of vitamin D in the body is produced in the skin. Consequently, the more clothing you wear, the darker your skin, the further away you live from the equator, and the greater the time spent indoors, the greater your risk of deficiency. For example, those who live north of

Birmingham in the United Kingdom, won't be able to synthesize vitamin D from October to March because of diminished exposure due to the low angle of the sun. Everyone in the United Kingdom is in danger of vitamin D deficiency for at least three months during winter. Anyone living above or below 34° latitude is at risk of experiencing some degree of vitamin D depletion during winter months.

Best estimates are that at around 40° latitude during a sunny summer day, a fair-skinned person could achieve significant vitamin D3 production in 10 minutes of exposure to the face and forearms to noon sunlight several times a week. The time needs to be increased to 30 minutes for darker skinned individuals. In people of the same age and skin color, there is considerable variation in serum levels of vitamin D with similar levels of sun exposure or supplements.

Under ideal conditions, 30 minutes of full-body exposure of pale skin to strong sunlight can result in the synthesis of between 10,000 and 20,000 IU of vitamin D. Such large amounts are more than capable of supplying the full needs of the body. The overproduction of vitamin D in direct sunlight does not happen because excess intermediate molecules are converted to inactive compounds by the very same sunlight used to make them. The synthesis of vitamin D starts to fade after five to 10 minutes, depending on skin pigmentation, the intensity of UVB radiation, and the amount of D3 precursors. After an initial burst of exposure to UVB rays, further exposure is of little benefit, but is harmful as it increases the risk of skin cancer and aging. It's therefore more productive and safer to regulate exposure to direct sunlight to several short bursts rather than a single extended period.

14.14 Vitamin D and sunscreens

While sunscreens help to protect against skin cancer and aging, they limit vitamin D production, causing a greater risk of other cancers. Depending on your vitamin D status and skin color, you should consider applying sunscreen after the first 15 minutes or so of exposure to strong sunlight.

15 Viruses

Following tobacco use, microbial infections are the most preventable cause of cancer in humans. An understanding of how pathogens cause cancer provides insights into carcinogenesis, in general, and ways of countering infections, in particular. It is a far greater challenge to successfully treat cancer once it has taken hold than to reduce infections through better screening, sanitation, and vaccination programs. In countries where infections from cancer-causing microbes are prevalent, there is an opportunity to make a significant contribution to the reduction of cancer incidence by tackling infection rates.

As early as the nineteenth century, it was noted that cervical cancer patterns corresponded to those of sexually transmitted diseases. In 1926, the first Nobel Prize for physiology or medicine in the field of cancer research was awarded to Johannes Fibiger, a Danish scientist, for his claim that a parasite caused stomach cancer in rats. Later, this was found to be untrue, a reflection of how far off the mark we were at the time. Significant progress in our understanding of the role of pathogens in the development of cancer didn't take place until the 1960s, quite some time after the invention of the electron microscope in 1931, which enabled us to see them.

Peyton Rous is credited as the first person to show that cancer can be induced by infection. In 1911, he injected a cell-free extract prepared from the sarcoma of a chicken into another chicken, which subsequently developed a sarcoma. Although at first his work was not generally recognized, the Rous sarcoma virus was named after its discoverer, who was awarded a Nobel Prize in 1966, some 55 years later. Whereas the development of a sarcoma in the second chicken was clearly attributable to a viral infection, such a distinct association is not readily assessed in humans.

Tracing a specific cancer back to an infectious cause is a difficult task that requires biological evidence by way of viral DNA or RNA present in the host, supported by epidemiological data. Spurious associations may arise due to complicating factors such as:

- The high prevalence of pathogens in the general population
- The length of time between cause (the infection) and effect (the cancer)
- The relevance of other determinant factors

Twenty-six mammalian viruses were associated with cancer between 1951 and 1972, two of which infect humans:

- The hepatitis B virus, which was discovered by Baruch Blumberg in 1963
- The Epstein-Barr, which virus discovered by Epstein and his colleagues in 1964

A list of viruses currently implicated in the development of cancer in humans is presented Table 15.1. For his work, the American physician Baruch Blumberg was awarded the Noble prize in physiology in 1976. The 1960s marked the start of an era when viral infection as a cause of cancer went viral. Also viral was the ephemeral "flower power" movement; the mantra to "make love, not war," and the anthem "Give Peace a Chance" by John Lennon. Whatever happened to "Give Peace a Chance"? As Bob Dylan would say, "the answer my friend is blowin' in the wind/the answer is blowin' in the wind."

All the crimes committed, day by day
No-one tries to stop it in any way
All the peace makers turned war officers
Hear when I say

"Police and Thieves" by Junior Murvin

Infectious organisms suspected of causing cancer share a few broad characteristics:

- They are often highly prevalent within the host population.
- They are present for long periods of time in an infected host.
- They are asymptomatic.
- Only a small percentage of infected hosts develop cancer.

Table 15.1 Cancer-Causing Viruses

Virus	Genome	Cancer
Helicobacter pylori	DNA	Stomach cancer
Hepatitis B	DNA	Hepatocellular carcinoma
Hepatitis C	RNA	Hepatocellular carcinoma
Human papillomavirus	DNA	Cancers of the cervix, vulva, anus, penis, and throat
HIV	DNA	Burkitt's lymphoma, Kaposi's sarcoma, cervical cancer

We can safely assume an infection on its own is not sufficient to cause cancer in humans, and other determinant factors are required.

15.1 Communicable diseases

In current times, excluding AIDS, about 20% of cancer deaths worldwide occur as a result of communicable diseases. The figure is nearer 7% in developed countries. Long-term infections of the bacteria *Helicobacter pylori*, infections of hepatitis B and C viruses, and infections of the human papillomavirus can largely account for the 20% mortality figure:

- Helicobacter pylori, stomach cancer, 9%
- Hepatitis B and C, liver carcinomas, 6%
- Human papillomavirus, cervical carcinomas, 5%

15.2 Human immunodeficiency virus

Infections of the human immunodeficiency virus (HIV) can progressively develop into full-blown AIDS, a disease in which the immune system is attacked and suppressed. Sufferers of AIDS demonstrate an increased risk of developing cancers such as Burkitt's lymphoma, Kaposi's sarcoma, primary central nervous system lymphoma, and cervical cancer. Kaposi's sarcoma is the most common cancer, occurring in 10%–20% of HIV victims. The second most common cancer is lymphoma, which is the cause of death for nearly 16% of AIDS sufferers. Both of these cancers are associated with the human herpes virus. Cervical cancer occurs more frequently in those with AIDS due to its association with the human papilloma virus.

15.3 How pathogens drive cancer

By what means do pathogens cause cancer?

- First, cells infected with pathogens are killed by the immune system as a means of combating the disease. This increases cell turnover, which increases mutation rates, and in turn promotes cancer.
- Second, long-term, persistent infections induce chronic inflammation that forms part of an immune response. This increases the production of oxidizing agents used by killer cells to terminate pathogens and cells infected with them. The fallout from this is damage to molecules, such as DNA, proteins and cell membranes, all of which may deregulate cell function and promote tumor formation. Although no single causative mechanism has been identified, increased oxidative stress has the potential to cause the accumulation of mutations, and may also promote fibrosis. In

addition, chronic oxidative stress may rewire cellular pathways, thereby contributing to cellular transformation.
- Third, viruses corrupt the DNA of their hosts in several detrimental ways, such as the introduction of foreign oncogenes, the deletion of host DNA, the translocation of chromosomes, and the amplification of host and viral DNA.
- Fourth, pathogens produce substances that may deregulate cellular processes, such as the cell cycle, DNA repair, cell death, and the immune response.
- Fifth, infections by pathogens weaken the immune system, which is important in fighting off new infections, as well as suppressing the growth and mobilization of cancer cells.

15.4 *Helicobacter pylori*

Long-term infections of *Helicobacter pylori* have been associated with gastritis, peptic ulcers, and gastric cancer, the second most prevalent type of cancer globally. It is commonly spread by contact with infected saliva or by fecal contamination of food or water. The WHO has classified *H. pylori* as a carcinogen. Unlike most other bacterial infections, which are duly eliminated by the immune system, the bacterium may establish itself for decades, or even the lifetime of its host, despite vigorous responses by both innate and adaptive immune systems. These defenses not only fail to clear the infection, but they also appear to assist its residence. In addition, we have seen cancer cells recruit immune cells to create a more hospitable environment. Reinfection commonly occurs following elimination of *H. pylori* in regions of the world with high incidence rates, which suggests a protective immune response is rare. It is possible that in some hosts, the immune response manages to clear Helicobacter pylori from the stomach. The frequency at which this occurs is unknown.

H. pylori has been infecting humans for over 30,000 years, which has given it time to evolve multiple ways of evading our immune systems, and ways of coping with the harsh acidic conditions of the stomach. The mechanisms whereby it can colonize more than half of the global human population, and persist for decades, are not clearly understood. A host of innovations that enable *H. pylori* to survive have been proposed. Interestingly, a few of these also are used as survival tools by cancer cells:

- The evolution of flagella providing motility that allows it to penetrate the mucus layer, and escape the high acidic conditions of the intestinal lumen.
- The production of the enzyme urease that catalyzes the formation of alkaline ammonium ions from urea, which makes its immediate surroundings less acidic and more habitable.

- The production of outer membrane proteins that permit it to adhere to gastric epithelial cells, and thereafter influence host responses.
- The avoidance of detection by the immune system by binding host proteins that form a defensive shield.
- Variation of its surface molecules to avoid detection by the immune system.
- The production of molecules that interact directly with immune cells and disarm their ability to kill.

An empty stomach is very acidic, which prevents the proliferation of most bacteria in the gastric lumen. *H. pylori* avoids this by living between the protective mucus layer that coats the stomach and the epithelium cells underneath, where the environment is less acidic. It makes this area even more habitable by using urease to form ammonium ions that increase its pH. It has been observed that strains of *H. pylori* defective in the production of urease or flagella are unable to colonize animal models.

Most infections of *H. pylori* occur during childhood. In developed countries, it has largely been eliminated due to the common practice of administering antibiotics to children, and due to better sanitation. Its prevalence varies widely between countries, with high rates in eastern Asia, Eastern Europe, and parts of South America. The sheer length of many *H. pylori* infections is a major contributor to its carcinogenic effect.

H. pylori is primarily a noninvasive organism, but it can attach to the surface of epithelial cells, thereby gaining access. This also permits delivery of toxins and effector molecules into cells. Chronic infection by *H. pylori* is considered to increase the stomach cancer risk up to six-fold. Persistent inflammation over decades is the likely cause. This response involves the attraction of killer cells, with consequent increased release of oxidizing agents, which induce damage to the DNA, proteins, and membranes of gastric cells. The consequent increased cell turnover and degradation of the protective mucous barrier are drivers that promote cancer development. The observed regression of precancerous lesions following the use of antibiotics also implicates persistent inflammation as a causative factor. The development of a vaccine against *H. pylori* is proving difficult.

15.5 Hepatitis B

Both hepatitis B and C viruses cause liver cancer, the third leading cause of cancer deaths worldwide. Although incidence rates have been stable, or in decline for many types of cancers, they have increased substantially for hepatocellular carcinoma in recent years, notably in Japan and the United States.

Hepatitis means inflammation of the liver. Many of those suffering from chronic hepatitis B virus infection are not aware, since they do not feel or look sick. Chronic infection is a major global health problem that puts sufferers at high risk of death from liver cirrhosis and cancer. The overall relative risk for hepatocellular carcinoma is estimated to be 13.7 times higher for chronic sufferers.

Infection of the hepatitis B virus is prevalent in sub-Saharan Africa, East Asia, the Middle East, southern parts of eastern and central Europe, and the Indian subcontinent. Chronic rates vary from 1% in Western Europe and North America to 10% in Saharan Africa and East Asia. Hepatitis B is estimated to be responsible for more than half of all liver cancer cases globally. The likelihood of an infection becoming chronic varies with age of infection:

- Infants under one-year-old, 80%–90%
- Children under the age of six, 30%–50%
- Adults, less than 5%

It has been estimated that 20%–30% of adults who become chronically infected with hepatitis B go on to develop liver cirrhosis and cancer.

Hepatitis B is spread when blood, semen, or other body fluid infected with the virus enters the body of an uninfected person. The most common route of infection is from mother to infant. Sexual transmission of hepatitis B also may occur, particularly in persons with multiple sex partners, and unvaccinated men who have sex with other men. Transmission of the virus also may occur through the reuse of infected needles and syringes.

On infection, the hepatitis B virus integrates its DNA with host DNA that compromises its integrity and promotes mutations. Common structural alterations include insertions, deletions, translocations, and duplications. Viral DNA does not appear to be a main driver of liver carcinogenesis as DNA fragments of hepatitis B are absent in the DNA of tumors. It is therefore likely, the cancer promoting effects of hepatitis B are through host mutations.

A vaccine that protects against hepatitis B virus infection has been available since 1982. It is 95% effective in preventing infection and chronic liver disease. For example, vaccination against chronic hepatitis B infection in infants in Taiwan has been very successful in dramatically reducing the number of children that develop liver cancer.

15.6 Hepatitis C

Hepatitis C is a liver disease caused by infection of the hepatitis C virus. This virus is an RNA virus that replicates in the cytoplasm of host cells and is not incorporated into its DNA. Most of those infected with the hepatitis C virus fail to clear it and are at long-term risk of progressive hepatic cirrhosis and hepatocellular carcinomas. Most hosts to the hepatitis C virus do not display symptoms of infection until liver damage becomes apparent. A significant number of those who are chronically infected go on to develop liver cirrhosis or cancer.

Hepatitis C is found worldwide, with higher rates in Africa, Central Asia, and East Asia. Globally, between 130 and 150 million people have chronic hepatitis C infections. About 3.5 million people in the United States have the disease compared to around 215,000 in the United Kingdom There are multiple strains of the hepatitis C virus, the distribution of which varies according to region.

Hepatitis C usually spreads when blood from a person infected with the virus gains entry into the body of an uninfected person. The most common modes of infection of the hepatitis C virus are via the use of infected injection needles, poor sterilization of medical equipment, and blood transfusion. Hepatitis C is not transmitted by airborne means or by saliva. Sharing a cup of tea or kissing are deemed to be safe. On rare occasions, hepatitis C may be transmitted via sexual intercourse.

The development of hepatocellular carcinoma is driven by two different means:

- Indirectly by chronic inflammation with associated oxidative stress and consequent cellular damage
- Directly by virus-specific effector molecules that interfere with cellular processes of the host

The exact mechanics of both means are still being worked out. The action of killer cells of the immune system and virus-induced inflammatory responses are likely to result in repeated cycles of the destruction and replacement of liver cells, which is highly conducive to the development of cancer.

Cirrhosis is a major risk factor for hepatocellular carcinomas, independent of viral infections. Persistent hepatitis C virus infections are typically associated with inflammatory and wound- healing responses within the liver. These mechanisms drive the synthesis of fibrous tissue, deposition of extracellular matrix proteins, scarring, and ultimately cirrhosis. The majority of hepatitis C virus-associated hepatocellular carcinomas develop in a cirrhosis setting. However, the formation of carcinomas in the absence of cirrhosis demonstrates it is not a prerequisite for cancer.

There is currently no vaccine for hepatitis C, and antiviral therapies are ineffective against many patients with chronic hepatitis C.

15.7 Human papillomavirus

There are more than 100 types of the human papillomavirus, only a small number of which have been implicated in causing cancer. Its prevalence in developed countries is approximately 7%, compared to 15% in developing countries. Infections of the virus usually clear without treatment within a few months. After two years, 90% are cleared, but repeated infections may occur. Human papillomaviruses are tissue specific and have primarily been associated with cancers of the cervix, the vulva, the anus, the penis, and throat. Genital warts, which are highly infectious, are also very common. Nearly all cases of cervical cancer can be attributable to human papillomavirus infections. Following exposure, it may take between 10–20 years for it to develop. Symptoms of cervical cancer tend to appear only after the disease has reached an advanced stage, and include abnormal vaginal bleeding after sexual intercourse, fatigue, weight loss, and vaginal discharge. On a global scale, an estimated 471,000 new cases of cervical cancer are diagnosed each year, 80% of which occur in less developed countries.

Human papillomavirus is transmitted via penetrative sex, genital contact, and, quite possibly, oral sex. The jury is still out on kissing. Worldwide, human papillomavirus is the most common sexually transmitted disease in adults. The virus is more prevalent in those who were sexually active at an early age and promiscuous. Rates of infection have been found to be lower in those who have never had it so good, and in those who have never had it. For instance, incidence rates are lower in nuns. Around 25% of mouth and 35% of throat cancers are human papillomavirus related. The types of human papillomavirus found in the mouth are almost entirely sexually transmitted, so it's likely that oral sex is the primary cause of infection. For the aficionados, occurrence rates suggest performing cunnilingus is riskier that performing fellatio.

Public health bodies in Europe, the United States, and Canada, as well as the WHO, recommend vaccination of young women against human papillomavirus to reduce the risk of cervical cancer and genital warts. In the United Kingdom, before they become sexually active, girls at the age of 12–13 are offered the human papillomavirus vaccine. The aim of this endeavor is to prevent seven out of 10 cervical cancers from developing following infection. The vaccine used gives protection for at least 20 years. More than 90% of human papillomavirus-related oropharyngeal cancers are caused by HPV-16, a particularly dangerous strain and the main cause of cervical cancer. Two vaccines are used to prevent cervical cancer, both of which protect against HPV-16.

In the United Kingdom, the number of deaths from cervical cancer has decreased by more than 40% during the past 20 years, a commendable achievement attributable to vaccinations and screening programs. Early treatment can prevent up to 80% of cervical cancers. In contrast, rates of throat cancers are rising, and set to increase further. A growing proportion of cases have been linked to the human papillomavirus. It has been shown that virus-positive and virus-negative oropharyngeal cancers have completely different risk profiles. Those with human papillomavirus-positive cancer tend to have had many oral-sex partners, whereas those with human papillomavirus-negative cancers tend to be heavy drinkers and cigarette smokers.

There are two types of tests that can be used to screen for early signs of cervical cancer:

- A Papanicolaou (Pap) test that detects cells early in the process of transformation
- A test for the human papillomavirus

The American Cancer Society recommends:

- All women should begin cervical cancer screening at the age of 21.
- Women between the ages of 21 and 29 should have a Pap test every three years.
- Women between the ages of 30 and 65 should have both a Pap test and a human papillomavirus test every five years, or alternatively a Pap test alone every three years.
- Women over the age of 65 who have had regular screenings with normal results should not be screened for cervical cancer.
- Women who have been diagnosed with cervical pre-cancer should continue to be screened.

In the United Kingdom, the National Health Service invites women for cervical screening as follows:

- Every three years for women aged 25–49
- Every five years for women aged 50–64

Women over the age of 65 are only screened if they haven't had a test since the age of 50, or if they have recently had abnormal tests.

It appears human papillomavirus causes throat cancers much in the same way it causes cancers of the cervix. The incorporation of viral DNA into host DNA appears to be a key event since cancer is rare in its absence. Such incorporation triggers the synthesis of two harmful proteins E6 and E7, which bind to, and shut down, two important tumor suppressor proteins, p53 and pRb. With p53 out of action, cell with damaged DNA multiply, which facilitates the accumulation of mutations that drive tumor formation.

In the absence of pRb activity, which inhibits cell cycle progression, cell proliferation continues unchecked.

It is becoming clearer that other bacteria may play a role in increasing or decreasing a person's susceptibility to cancer. A study of 70,000 individuals showed that patients with inflamed and bleeding gums due to poor oral hygiene had double the risk of cancers of the oral cavity and digestive tract. The risk of cancer rose with increasing severity of periodontitis and was specifically associated with the oral bacterium Porphyromonas gingivalis.

16
Metastasis

Cells that make up our tissues are regulated, so that they stay where they belong and obey boundary rules. If they were permitted to wander around en masse, taking selfies among new friends in new locations, pretty soon a mess would be created, tissue function would cease, and disastrous consequences would follow. Metastasis is a quintessential case in point. The inappropriate migration of cells to other locations, which depletes local reserves and disrupts the function of other tissues, is something that is tightly controlled. How then does metastasis occur, and how do we stop it from happening?

> *"Relax," said the night man,*
> *"We are programmed to receive.*
> *You can check-out any time you like,*
> *But you can never leave!"*
>
> "Hotel California" by The Eagles

Metastasis refers to the spreading of cancer cells from their tissue of origin, their primary site, to other distant tissues from where they may develop into secondary tumors. Most of the cells that successfully manage to migrate to new sites undergo cell death. Some may enter a state of dormancy, and a few go on to proliferate in the new microenvironment. The few that grow give rise to micrometastases, which eventually develop into macrometastases. At the time of diagnosis, at least half of cancer patients have detectable metastatic tumors. A greater proportion also carry micrometastases that are not detectable by conventional means.

16.1 Metastatic beginnings

There is some uncertainty regarding the point during tumor development at which metastasis begins. One school of thought suggests a linear progression model, which considers tumor cell dissemination begins after significant growth has taken place, while another school of thought suggests a parallel progression model, which proposes tumor cell dissemination commences early on, when the sizes of tumors are between 1 and 4 mm.

Under the linear progression model, the premise is that metastasis results as a consequence of the general genetic upheaval progression, which takes place in transformed cells and is preceded by benign and invasive stages. This is consistent with the general late establishment of metastatic tumors.

Under the parallel model, the supposition is that metastasis is triggered by one or more specific early genetic events. There is some evidence supporting this model. For example, the early formation of micrometastasis in the bone marrow and lungs of patients have been observed for HER2 induced breast tumors prior to the invasion of surrounding tissue. Numerous other studies show that circulating tumor cells gather at lymph nodes or bone marrow prior to the establishment of metastatic sites. In the linear progression model, the late establishment of macrometastases may be explained by the dormancy of newly settled cancer cells. This is in part due to their inability to immediately coerce surrounding cells in their new environment to assist in their development. At some point, which could be many years later, mutations may be acquired that complete the settling in process, after which growth begins.

A linear progression model makes it imperative that cancers, which are known to metastasize, are treated as early as possible, before they spread and become near impossible to cure. It also creates the opportunity for early diagnosis of cancers by identifying and classifying tumor cells circulating in the blood.

16.2 Metastatic sites

For reasons that are not entirely clear, some types of tumors are highly likely to metastasize, while others never do, and some only do on rare occasions. For example, melanomas that invade the tissue underlying the skin frequently metastasize, whereas basal cell carcinomas of the skin very rarely do. Cancers that spread to distant sites commonly originate from the lungs, breasts, the colon, the kidneys, the prostate, the pancreas, and the liver.

As is evident from Table 16.1, primary tumors metastasize preferentially to a limited number of secondary sites. For example, breast cancers tend to

Table 16.1 Metastatic Sites

Primary Site	Likely Metastatic Sites
Bladder	Brain
Breast adenocarcinoma	Brain, bone, lungs, liver
Colorectal	Lungs, liver, lymph nodes
Kidney clear cell carcinoma	Bone, liver, thyroid
Lung (non-small)	Brain, bone, liver, adrenal glands
Pancreas	Lungs, liver
Prostate	Bone
Skin melanoma	Brain, liver, bowel
Stomach	Esophagus, lungs, liver

metastasize to the brain, bone, lungs, and liver, while prostate cancers favor the bone.

As circulating tumor cells flow thorough the blood system, the vessels they encounter become narrower and narrower. The internal diameter of capillary vessels is 3–8 μm compared to 20 μm for an average cancer cell. A point is reached when circulating cells become enmeshed in blood vessels that are too narrow for them to travel any further. This is most likely downstream from their point of entry into the blood circulatory system. The lungs and liver are the two most common sites where they end up. This is for two reasons. First, blood leaving many organs flows to the lungs to be oxygenated. Therefore, the likelihood of metastasis is relatively high. Second, from the intestines blood travels to the liver, so cancer cells of the intestines tend to metastasize there. Some cancer cells are able to settle on sites distant from their primary site. For example, prostate cancer cells spread to bone before spreading to other organs. This selectivity may be due to preferential adhesion of prostate tumor cells to bone tissue.

Once a cancer spreads to other locations, it may be difficult to determine its site of origin. When this happens the growth is referred to as a cancer of unknown primary. Up to 5% of metastatic cancers belong to this category.

16.3 Cell adhesion

Cell adhesion is something you've probably never gotten excited about, but if it wasn't for it, you would simply fall apart. There are cells, such as brain cells, that function in stationary mode, while there are cells, such as blood platelets, that are constantly on the move. There are biological processes, such as wound healing, for which some stationary cells need to move, and some moving cells need to become stationary. To repair a wound, moving blood platelets need to form clots onsite to stop bleeding, and stationary cells need to move to the site of damage to partake in the healing process. Therefore, cell adhesion is something that needs to be turned on and off as and when required.

The cells of any given tissue are tightly bound to each other and the extracellular matrix, which is the non-cellular component of tissues that fills the gap between cells. Such binding is facilitated by calcium and magnesium ions. The extracellular matrix is important for tissues and organs to attain their correct form, and serve their correct function. It is composed of fibrous protein molecules and polysaccharide molecules secreted into intercellular spaces that provide scaffolding for cell adherence.

During development, in addition to acquiring a particular fate, cells in a tissue must adopt the correct form. Tissues take shape in many ways that are influenced by changes in cell adhesion, shape, and migration. Cell form also is provided by the cytoskeleton inside cells, which is made up of filamentous proteins, and provides mechanical support to cells and their cytoplasmic constituents.

The adhesion of cells to the extracellular matrix is primarily via proteins that span the outer cell membrane, and are collectively referred to as integrins, as shown in Figure 16.1. They also are involved in cell-to-cell binding and the passing of signals between extracellular sources and the inside of cells. Integrins constitute a large family of proteins that are the principal receptors of cells for binding to proteins of the extracellular matrix. They are very important because they are the main means by which cells respond to the extracellular matrix. Integrins differ from other receptors in that they typically bind with lower affinity and are 10 to 100 times more abundant on the cell surface. On occasions when cells need to migrate, weak integrin bonds are readily broken to permit movement.

Integrins may be in an active state in which they are able to bind, or in an inactive state in which they are not. Activation is brought about by a change in conformation of the molecule part that sticks out of the outer cell membrane, brought about by modifications to the part of the molecule on the inside of cells. In normal stationary cells, the default setting for integrins is the active, adhesive state, while the default setting for cells constantly on the move is an inactive, non-adhesive state.

Freely circulating cells such as blood platelets and white blood cells need to adhere when called into action. For example, upon activation by contact with damaged blood vessels, platelets in the blood become adhesive and form clots by binding to fibrinogen, a soluble protein. Their ability to adhere is

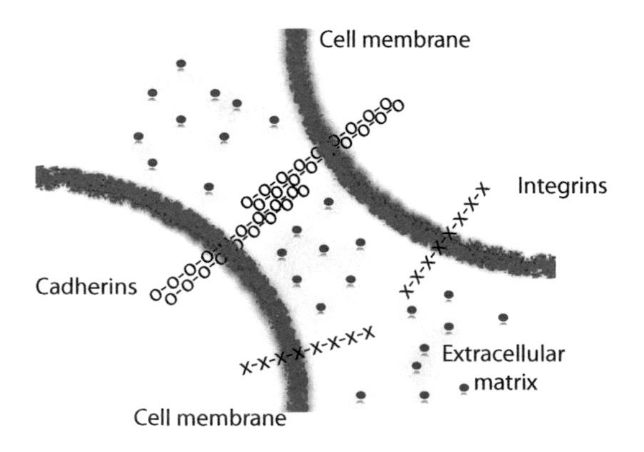

Figure 16.1 Cell–cell adhesion by cadherins and cellextracellular matrix binding by integrins.

turned on by modifications to the structure of integrins on the inside of cells. When T lymphocytes bind to antigens on the surface of antigen-presenting cells, intracellular signaling pathways are triggered that activate integrins. This permits T-cells to adhere more strongly to the antigen-presenting cells, so that they remain in contact for an extended period of time. Inactivation of integrins allows T-cells to disengage and return to a non-binding state.

As the extracellular matrix of each tissue type is unique, a cancer cell that has moved to a new location has to adapt to an environment that is likely to be very different to the one it came from. This is one of the reasons why the vast majority of cells that escape from cancer colonies do not develop into tumors at distant sites, and also why there is selectivity in the sites where they are able to develop.

How are cells bonded together? Calcium dependent adhesion molecules, otherwise known as cadherins, are a superfamily of proteins that span the membrane of cells into the cytoplasm. More than 20 types have been identified that are involved in maintaining tight adhesion between cells, which is calcium dependent. Cadherins are evolutionary ancient and expressed by the cells of all vertebrates. The main component of contact points between epithelial cells is E-cadherin. As shown in Figure 16.1, this helps them to stick together, and form bridges that connect their insides. Evidence suggests that they play more than a passing role in tumor progression. If a site of tissue invasion was a crime scene, the fingerprints of cadherins and integrins would be all over it.

Cadherins exhibit a tissue-specific pattern of expression, for example E-cadherin, M-cadherin, and N-cadherin are expressed in epithelium cells, muscle cells, and mesenchymal cells, respectively. Cells of an epithelial phenotype are present in skin and line cavities of body organs such as the digestive system, breast and lungs. Cells of a mesenchymal phenotype are found in tissues such as connective tissue, lymphatic tissue, and blood vessels. Cadherin complexes form at contact points between cells and play an essential function in the regulation of cell adhesion and in cell-to-cell communication. The cytoplasmic section of cadherins interacts with molecules inside cells such as catenins that influence cell adhesion. Mutations in either E-cadherin or catenins have been observed to reduce cellular adhesion.

A loss of E-cadherin has been observed in several types of cancer. In an experimental tumor model, administration of it could block the ability of cancer cells to spread, which has made its use for therapeutic purposes a possibility.

16.4 The metastatic cascade

"A journey of a thousand miles begins with a single step" is a saying attributed to Chinese philosopher Lao Tzu over 2500 years ago. Metastatic cancer cells must have been the furthest thing from his mind at the time, but nevertheless it applies to them. The first step cancer cells must take when moving out is to dislodge themselves from neighboring cells and the extracellular matrix to which they are attached. Metastasis follows a sequence of events, referred to as the metastatic cascade. Getting cells out of a comfort zone to up and leave should be as difficult as getting young adults that you have raised to leave home. Unlike the five stages they go through denial, anger, depression, bargaining, and acceptance, cancer cells that relocate must:

- Invade surrounding tissue
- Detach themselves from the tumor mass
- Migrate to a nearby circulatory system
- Intravasate by boring into the blood or lymphatic system
- Escape the attention of circulating immune killer cells
- Survive abrasive damage from being squirted at speed around the body
- Survive chemical attacks from reactive agents in circulation
- Extravasate by boring out of the blood or lymphatic system
- Settle at a new tissue site
- Flourish in a new and unfamiliar setting

An illustration of the metastatic process in presented in Figure 16.2. Tumor cells unable to complete all the steps of the metastatic cascade do not form secondary growths.

In a colony of tumor cells consisting of many different sub-clones, only a subset will have the full complement of mutations to be able to separate out, migrate to blood vessels, and intravasate. Not all of those that manage these feats will be equipped to survive the journey to another area of the body, and once there, not all will be able to commence growth, immediately or at some point in the future. If cells that have settled can survive indefinitely, they may in time evolve the means of growing in a

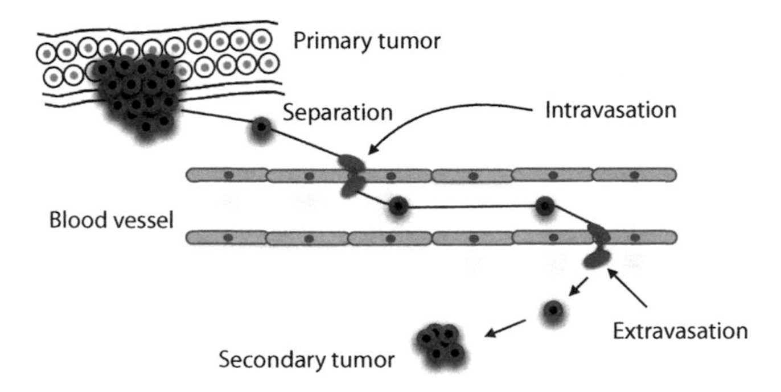

Figure 16.2 Metastasis.

new environment. Given the barriers a migrating cell has to negotiate, the remarkable thing about metastasis is that it happens at all.

Circulating tumor cells have been found in the blood of many cancer patients. It appears more than a million cells can be shed each day from one gram of tumor, which is made up of approximately one billion cancer cells. Metastasis appears to be a numbers game in which wave after wave of cells are unleashed, out of which only a few survive, and an unknown number go on to form new tumors. With such high numbers of candidate cells, and such high rates of attrition, as a consequence of natural selection, only the fittest cells survive. This stern filtering process ultimately preferentially selects metastatic cancer cells more robust and adaptable than those of the original tumor, which suggests they may be more difficult to kill. This appears to be the case.

16.5 Invasion

Benign tumors cells that do not venture out of their local comfort zone do not metastasize. For them to spread to distant sites, they need to acquire the ability to move toward blood or lymph vessels and penetrate them. To accomplish this, cells need to cut paths through the extracellular matrix, the basement membrane that forms a barrier between tissue types, and the basement membrane that surrounds capillary vessels. A basement membrane is a thin, fibrous, extracellular matrix of tissue that separates different tissue types.

When cancer cells move, they do so collectively as a unit, as detached clusters, or as single cells. To move, tumor cells require assistance from their microenvironment, which provides the necessary signals, and which helps in clearing paths for migration. The migration of cells involves the loss of cell-to-cell adhesion, and the expression of a set of internal proteins that facilitate cell motility. This is accompanied by a transition of cell phenotype from epithelial to mesenchymal. Mesenchymal to amoeboid phenotype transitions have been observed. During cancer progression, many tumor cells undergo phenotype conversions. For tumor cells to become motile, protrusions are formed to extend forward, followed by adhesion to surroundings, and then by trailing end contraction. These synchronized events drive cells forward.

Cells that become free from their neighbors need to forge a route through the dense extracellular matrix. For this purpose, normal neighboring cells are recruited to help. They are induced to release enzymes such as metalloproteinases that are generally produced in an inactive form, and which are activated by proteins produced by cancer cells. Metalloproteinases digest basement membranes and the extracellular matrix around tumors, thus clearing a path and allowing individual cells, or clumps of cells, to migrate.

Macrophages have been observed to be associated with tumor cells on the move, raising suspicion about the alliance of our immune killer cells, whose duties are to serve and protect. Such tumor-associated macrophages have been classed public enemy number one under suspicion of aiding and abetting. Evidence of their complicity is as follows:

- They loiter with intent at sites of tumor invasion.
- They are often found at sites where cancer cells intravasate into blood or lymphatic vessels.
- They have an armory of potions such as cytokines, growth factors, and proteolytic enzymes, all of which are known to enhance both tumor movement and metastases.

Proteolytic enzymes cleave bonds of protein molecules, thereby loosening cell-to-cell contacts and cell to extracellular matrix contact. They play an important role at different stages of the metastatic cascade, including invasion and intravasation. Macrophages are the major contributors of proteolytic enzymes to the tumor microenvironment.

Long ago, the circumstantial evidence given above would have been sufficient for tumor-associated macrophages to be found guilty of witchcraft and burned at the stake for their sins. There is more:

- There are correlations between macrophage density and poor patient prognosis. For example, the accumulation of tumor-associated macrophages in breast carcinomas has been unequivocally correlated with a poor prognosis.
- Treatments that reduce the recruitment of macrophages by tumors also reduce tumor progression and enhance sensitivity to radiotherapy.

16.6 Detachment

What could possibly possess a tumor cell to leave an environment it is acquainted with to take on a suicide mission? The short answer is we don't know, but I'm willing to speculate. The blood vessels of tumors are not as well constructed as those of normal vessels and are more leaky. Their construction is driven by an unbalanced expression of a small number of growth factors in the local environment, in particular vascular endothelial growth factor A (VEGF-A). At least six distinct types of blood vessels are formed that differ from each other, and from normal blood vessels, with respect to organization, structure, and function. Whereas normal blood vessels are evenly branched in a hierarchal manner consisting of well-formed arteries, arterioles, capillaries, venules, and veins, tumor blood vessels are unevenly branched and inconsistent in their composition. They often are larger than normal vessels and generally have weak cell-to-cell junctions through which cancer cells can enter the blood circulatory system.

The supply of blood vessels is richer at interfaces, where tumor cells meet host cells, than in the center. The density of a blood vessel tends to decrease as tumors grow, leading to zones of low blood supply that ultimately lead to cell death, as tumors outgrow their blood supply. Cells that by chance are starved of nutrients, due to poor blood supply or space, cannot flourish and are in danger of dying via necrosis. Many fast-growing tumors have dead cells, presumably caused by a lack of nutrients and oxygen. Carcinoma patients with tumor necrosis are subject to less favorable outcomes. Unlike apoptosis, necrosis is usually associated with inflammation, which, when chronic, is associated with tumor development.

Metastasis may be initiated when a sub-clone of cells by chance manages to acquire mutations that revert their cell phenotype to one more suited to cell separation, migration, and intravasation. Alternatively, it could be initiated as a survival response to stress, triggered by a shortage of resources such as space, nourishment, or oxygen. The adoption of an exit strategy for stressed cells in an environment where long, dark days are followed by long, dark nights, and where leaky blood vessels provide a means of escape, makes sense. This could be a universal feature of life, necessary for survival that is programmed in our DNA.

> *There is a tide in the affairs of men, Which taken at the flood, leads on to fortune. Omitted, all the voyage of their life is bound in shallows and in miseries. On such a full sea are we now afloat. And we must take the current when it serves, or lose our ventures.*
>
> *Julius Caesar* by William Shakespeare

It is increasingly apparent that cell detachment from primary tumors involves an epithelial to mesenchymal transition that enhances their migratory and invasive capabilities.

In adults, the transition of cells from an epithelial type to a mesenchymal type is a normal occurrence, induced in response to wound healing and inflammation. During this process, epithelial cells undergo a series of structural changes that includes, adoption of a mesenchymal phenotype and detachment from one another. These are similar to the observed metastatic changes undergone by transformed epithelial cells in carcinomas. An early event in the transition from an epithelial cell type to a mesenchymal cell type is the cadherin switch. This involves a reduction in levels of E-cadherin, accompanied by an increase in levels of N-cadherin, a protein present in mesenchymal cells. N-cadherin destabilizes the adhesion of epithelial cells, thereby promoting cell separation.

16.7 Anoikis

A single cell assesses whether it is in the right location using receptors on its outer membrane that interact with surrounding cells and the extracellular matrix. When normal cells lose interactions with their surroundings, the cell cycle is arrested and a process, termed anoikis, triggers cell death. It is in effect apoptosis, induced by a lack of correct cell attachment. There are many diverse ways in which apoptosis is initiated through the passage of signals via adhesion receptors between cells. The appropriate signals from the extracellular matrix also are required to prevent the triggering of anoikis. The way a particular type of cell responds to incorrect adhesion may differ from that of another cell type. Each cell type possess its own unique collection of signaling proteins, wired in its own unique configuration.

Anoikis provides a means of ensuring tissue integrity, maintained by terminating cells that have become loose cannons. The disarming of an anoikis response is something an aspiring metastatic cell needs to achieve and preserve at each step along the way. A variety of strategies have been adopted. These involve the rewiring of complex signaling pathways.

Two mechanisms have been identified by which tumor cells may overcome anoikis:

- They may regain the ability to migrate and proliferate by rolling back differentiation to alter their cell phenotype, from an epithelial type to a mesenchymal type.
- They may reconfigure the array of receptors on their cell surface.

16.8 Intravasation

Approximately 80% of all cancers are carcinomas that originate from epithelial cells. These are present in skin and they also line cavities of organs of the body such as the digestive system, breast, and lungs. Epithelial cells form important protective barriers to pathogens and toxins, and so are tightly bound together. Therefore, they should provide a formidable barrier to the migration of single cells or clumps of cells.

Intravasation refers to the passing of cancer cells into the bloodstream or lymphatic system. To accomplish it, migrating cells need to traverse tissues between their primary tumor and carrier vessels, and squeeze through junctions between cells of blood vessels. Both intravasation and extravasation involve changes to the shape of a cancer cell and therefore, involve regulators that affect the cytoskeleton of cells. The blood system is considered to be the main route for metastasis, but there is increasing evidence it may be the lymphatic system. Whereas blood capillaries are covered by a basement membrane, lymphatic capillaries have thin walls with single layers of endothelial cells and junctions that are not as secure.

Although lymphatic vessels do not infiltrate tumors as blood vessels do, they provide a viable route for migration. From the lymphatic system, loose cells can readily gain access to the rest of the circulatory system and spread throughout the body. Once metastasis to the lymphatic system has occurred, the prognosis for cure drops significantly.

A blood circulatory system not only brings nutrients to tumor cells, it also serves as an escape route. The mechanism whereby intravasation occurs is poorly understood. It is not clear how macrophages and the tumor microenvironment contribute to intravasation, or whether tumor cells require abnormal blood vessels to intravasate. Macrophages are able to attract cancer cells toward blood vessels by secreting epidermal growth factor. In addition, they can secrete tumor necrosis factor, which loosens the junctions between endothelial cells and facilitates the entry of cancer cells into blood vessels.

Unlike red blood cells and white blood cells, circulating tumor cells from a stationary environment are not equipped to withstand physical stress. For the daunting journey, it's better to take some support by way of other cells, which is exactly what we believe happens. In addition, some circulating tumor cells rapidly acquire a cloak of adhering blood platelets, which bind to them and serve as shields that protect from direct contact with circulating killer cells in the blood. When this micro-coagulation is suppressed, the success rate of metastasizing cells in the development of new tumor colonies drops significantly. This fall is attributed to the prevention of attacks by killer cells, which would normally attack cancer cells out in the open. The acquiring of blood platelet cloaking technology by cancer cells that have managed to enter the blood stream is considered to be a routine occurrence. Cells on the inside of clumps are better protected from abrasive damage during circulation in the blood stream, and from attack by killer cells of the immune system.

16.9 Extravasation

Extravasation refers to the passing of cancer cells out of the bloodstream into surrounding tissue. This is a task killer cells of the immune system carry out on a regular basis, which means the genes that code for proteins that accomplishes it exist in circulating tumor cells. All that is required is the turning on of one or more biological processes.

Extravasation typically occurs in small capillaries in which cancer cells are trapped. To accomplish it, circulating tumor cells need to:

- Adhere to endothelial cells that line the inside of blood vessels
- Pass through junctions between endothelial cells or pass through them
- Traverse the basement membrane that surrounds blood vessels

Extravasation begins with the attachment of circulating tumor cells to endothelial cells that line vessels. It is apparent that multiple adhesion receptors on cancer cells contribute to their adhesion to endothelial cells, and that such receptors vary with cancer types. This may be facilitated by the adhesion of circulating blood cells to cancer cells triggered by the activation of integrins. Several integrins on cancer cells have been implicated in this process and in transmigration across blood vessels. Recruited blood cells may subsequently assist with extravasation. It is therefore likely that, in the treatment of cancer, several different types of receptors will need to be blocked to combat metastatic dissemination.

It's possible circulating tumor cells employ the same mechanism that killer cells of the immune system use to pass through blood vessels. Such cells routinely migrate in and out of the bloodstream to maintain surveillance, and to respond to stress calls made as a consequence of cell injury or infection. We know injured and infected tissues release a number of signaling molecules that increase the permeability of blood vessel, and induce endothelial cells to present adhesion molecules on their surfaces. Circulating white blood cells attach to these, and extravasate. Following attachment endothelial cells move apart allowing immune cells to squeeze through.

Like a Jedi mind trick, tumor cells may employ subterfuge to con their way through by presenting molecules on their surface that are similar to those of white blood cells. This mechanism offers another explanation to account for the selectivity of circulating tumor cells for certain tissue types, which may be homing in on sites that have sent out invitations via chemical messages. Proliferation has been observed to initially commence in the lumen of blood vessels of liver cells, after which they cross the endothelium and invade underlying tissues.

16.10 Treatment

The prevention of the dissemination and formation of new tumors needs to be a major treatment goal as 90% of cancer deaths occur as a consequence of metastasis. Currently, we have no proven means of stopping metastasis from happening. The choice of treatment generally depends on the type of primary cancer. For example, breast cancer that has spread to the lungs is still regarded as breast cancer, and is treated as metastatic breast cancer. Quiescent metastatic cells survive chemotherapy because they are not dividing. They also survive radiotherapy because they are far removed from the primary site under treatment.

Relapses that emerge long after primary tumors have been eliminated by surgery, chemotherapy, or radiotherapy are due in part to the sudden growth of metastatic colonies. These may undergo cell cycle arrest, or strike a balance between proliferation and cell death, and so lie dormant for periods of time, extending into decades at tissue sites foreign to their primary location.

About one in five metastatic breast cancer patients do not develop metastatic growth until 10 years after treatment. To commence growth, a metastatic cell needs its new environment to be supportive of its particular requirements, by responding positively to signals it releases. Cells may lie dormant because it takes time for them to acquire the ability to influence their microenvironment in order that it nurtures their growth. Inducing the development of an adequate blood supply is a prime example of this.

16.11 Are macrophages double agents?

As far as tumor progression is concerned, macrophages have a bad name due to their persistent presence in the tumor microenvironment, sleeping with the enemy. There is a fair amount of evidence that associates them with some of the hallmarks of tumors progression, such as angiogenesis and tissue invasion.

Macrophages that are found around tumors are normal, classified as M1. With the passage of time, as tumors progress, these are recruited to become part of a microenvironment complex customized to sustain it. The presence of macrophages causes inflammation, which among other things, promotes cell growth, blood vessel growth, and cell invasion. Enough to delight and titillate any colony of tumor cells. This is effected by the release of signaling proteins. In so doing, surrounding cells are prompted to assist the development of tumors. Macrophages that are led astray in this manner are classified as M2. What induces the conversion of M1 to M2 is currently unknown.

M2 macrophages may also promote local immunosuppression, thereby blocking other immune cells from launching attacks. There are suppressor cells that reduce the immune response once a threat has been eliminated. In some cancers they may be recruited to suppress attacks by the immune system.

16.12 Are neutrophils double agents?

A large body of clinical evidence suggests neutrophils are involved in cancer development. Their persistent and abundant presence at sites of chronic inflammation makes them guilty by association. Some tumors seem to actively recruit neutrophils by the release of chemokines that attract them. This may be promoted by the release of chemokines directly by tumor cells, or by prompting of immune cells to secrete them.

There is evidence that neutrophils influence tumor development through several other agents that are secreted. These include reactive oxygen species, enzymes that reduce cell adhesion, and agents that promote angiogenesis. Reactive oxygen species cause damage to DNA, which promotes cancer, while enzymes that reduce cell adhesion support tissue invasion and metastasis. Clinical studies indicate that neutrophil infiltration of tumors is associated with poorer prognosis. This association has been observed for solid tumors such as adenocarcinomas, cancers of the colon, hepatocellular carcinomas, melanomas, and non-small cell lung carcinomas. Neutrophils also may assist the migration of incipient cancer cells. It has been shown that neutrophils promote metastasis in cancers such as adenocarcinomas, breast cancer, melanomas, and skin squamous cell carcinomas.

17
Revelations

We have a much better understanding of how cancer is caused now than we did a mere 30 years ago. It is brought about by the accumulation of genetic aberrations that confer growth advantages to a cell over surrounding cells. These build up spontaneously and slowly over time, which may extend into decades, but their rate of accumulation is accelerated by many factors. Some genetic aberrations that drive tumor development are inherited, which is a key contributor to the development of cancer at a young age. The cells of those who carry such traits are already a step ahead on the road to malignancy. The risk of developing cancer is amplified because driver mutations present at birth exist in nearly every single cell of the body, making the possibility of one of them acquiring other mutations and becoming malignant very likely. The inheritance of mutations that impair the function of genes that regulate the cell cycle or repair DNA are particularly detrimental, because in the absence of their active contributions, mutations accumulate at a faster rate.

As a consequence of genetic modifications, oncogenes are switched on or over expressed, tumor suppressor genes are switched off or under expressed, and signaling pathways are rewired. The combination of genetic defects is unique for each tumor due to the stochastic way they accumulate, the large number of target driver genes, and the variety of ways in which genetic material can be modified. Mutations of gene-coding regions may give the proteins they code greater or lesser activity, while alterations of non-coding regions may raise or lower the amount of proteins expressed. The development of a tumor is illustrated in Figure 17.1.

17.1 Complexity of cancer phenotype

There are a number of factors that cascade and multiply the complexity of the cancer phenotype as we drill downwards from an individual to a single cancer cell:

- No two individuals are genetically alike.
- Within each individual, each cell type has its own unique gene expression profile.
- The signaling pathway of each cell type is wired differently.
- Each signaling pathway within a cell may be disrupted in multiple ways.

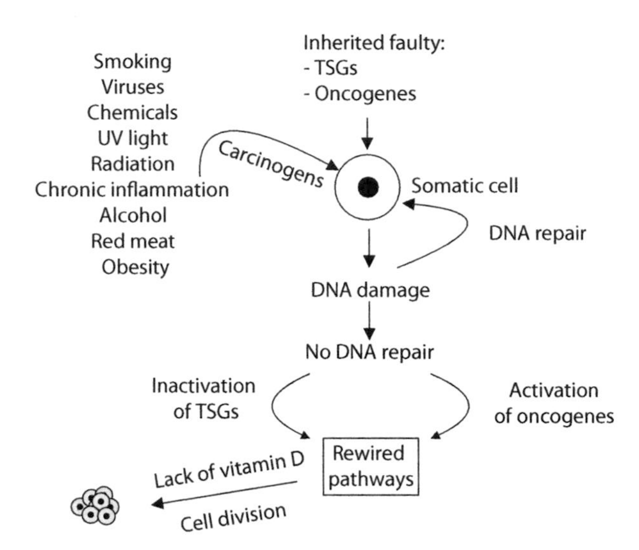

Figure 17.1 Development of a tumor.

- The genetic aberrations of tumors of the same cell type are different.
- Individual tumors consist of sub-clones of cells with different mutation profiles.
- Cancer cell DNA is unstable and new mutations are constantly happening.
- Chemotherapy and radiation treatment generate new mutations.

The fact that we are different means that some of us are more susceptible to certain types of cancer, and more resistant to others. It also means we may respond differently to similar treatment regimens.

The signaling pathways of each cell type are wired differently. Mutations that are drivers in one type of cell may not necessarily do so in other types. Thus, treatments that work for cancers of a particular cell type, may not work for cancers of a different cell type. Within a single colony of tumor cells, treatment that kills members of a specific sub-clone of a mutated cell may not necessarily be effective in killing members of other sub-clones. Following treatment, sub-clones that survive thrive in the absence of competition from other sub-clones that suppressed their growth prior to treatment.

17.2 Diagnosing cancer

The early detection of cancer is vital to successful treatment. Once it spreads to other sites, prognosis is poor. The accurate diagnosis of cancer type is critical for the determination of suitable treatment. For example, drugs such as Trastuzumab (Herceptin) and Imatinib (Glivec) that work against one type of cancer, may be ineffective against other forms.

At this point in time, there is no single test that can be used on all cancers to accurately identify type. Multiple tests may have to be taken, which slows down diagnosis and increases costs. Clinicians currently depend almost entirely on histopathology for their definitive diagnosis of solid tumors. Since the identification of these is primarily visual, there is room for error in diagnosis. This is because tumors that look alike under a microscope may have different genetic aberration profiles, and require different treatment. The light microscope, which was invented in 1590, has served us well, but it's time to move on. The process typically starts with patients noticing symptoms such as:

- The appearance of lumps
- The coughing-up of blood
- Loss of weight
- Chronic pain in bones
- Persistent headaches
- Change of bowel or bladder habits
- A sore that does not heal

In some cases, cancers are detected quite by accident during treatment of unrelated ailments.

A clinical evaluation begins with physical examination and history taking, by a general practitioner, then various screening or diagnostic tests for malignancy may be ordered. Based on these, a patient may be referred to a cancer specialist. The whole process may be costly and time consuming. As the chances of successful treatment diminish with each passing day, it is critical that the possibility of cancer is considered early, and an accurate diagnosis is made expeditiously.

17.3 Cancer types may also be identified by other techniques

Cancer type identifying techniques also include:

1. Testing for biomarkers
2. Imaging
3. Genetic testing

Together these approaches help doctors make better-informed decisions.

17.3.1 Testing for biomarkers

Biomarkers are molecules that are released from cancers, and are present in serum or urine. Many currently in use lack sufficient sensitivity or specificity to serve as a single test for accurate diagnosis. They are used to supplement diagnosis, in conjunction with other methods. The need may arise for several tests to be carried out, which adds time and expense.

17.3.2 Imaging

Imaging may be used for:

- Screening
- Diagnosis
- Treatment monitoring

Images of tumors and surrounding areas may be produced using ultrasound, x-ray, CT, CAT, or MRI scan. Mammography, which uses x-rays, has become routine for the screening of breast cancer. CAT scans are commonly used in cancer diagnosis, stage determination, and treatment monitoring. MRI technology is expensive.

17.3.3 Genetic testing

We are aware of the need to base the classification of cancers on their DNA mutation profiles. This has not been done to date, in part because we do not have a clear understanding of the full complement of genetic defects for each specific cancer. The use of gene mutation and gene expression profiles to distinguish between different forms of cancer present the holy grail of diagnosis. Progress is being made. Although we have the technology to identify defects in DNA, amidst numerous aberrations, we are currently unable to effectively identify causative DNA mutations, gene expressions, and chromosome abnormalities. The use of this technology as a prime method of diagnosing cancer type is currently a work in progress. Advantages of genetic testing would be:

- One tests for all cancers
- Faster diagnosis
- Greater accuracy of diagnosis
- A better classification system for cancer

The cost of whole genome sequencing is reaching a level at which routine analysis of tumor cells is affordable. The construction of gene expression profiles of tumors is currently carried out in many labs. Challenges to the establishment of genetic profiles as a means of diagnosing cancer include:

- The availability of funds
- The availability of skilled human resources
- The volume of work that needs to be done
- The standardization of data collection
- The computer processing power to manage the large volume of data

- The association of genetic aberration profiles with different forms of cancer
- The reclassification of tumors to DNA mutation types

There are ongoing global cancer genome sequencing projects gathering data that is laying a foundation that may be used for genetic testing. COSMIC, ICGC, and TCGA are prominent examples of such efforts.

17.4 Screening

There is a strong argument for the implementation of public health screening programs at a national level, which could save much suffering and loss of life. Some of these, such as Pap smears and breast x-rays, are already in place in some countries. Screening may be cost effective in the long run as patients struggle to overcome cancer for years, consuming expensive drugs and human resources in the process. There is a need for newer and more effective screening methods geared at early detection and treatment.

17.5 Lifestyle changes

There also is a need for prevention initiatives such as:

- A reduction in the number of cigarette smokers
- The control of alcohol consumption
- A reduction of the consumption of excessive amounts of red meat and processed foods
- A reduction in obesity of the population
- An increase in vitamin D intake
- Avoidance of long periods of exposure to sunlight
- Control of the intake of antioxidant supplements
- Avoidance of polluted air

17.6 Treatment

Prognosis described in terms of tumor shrinkage, five-year survival rates, and cancer-free, demonstrate how really inept we are at curing cancer. We don't talk in these terms so readily about other diseases. Following treatment, the outcome should be patient cured, and the possibility of relapses should be eliminated. Although chemotherapy and radiation treatment are currently being used with varying degrees of success, they cannot be the way forward. The use of chemotherapeutic drug cocktails is a refinement of a theme that should be phased out in favor of more effective treatment options, such as targeted therapy, or cocktails of targeted therapeutic drugs. There is a pressing need for alternate drugs that do not work purely by damaging dividing cells to the point where they commit suicide.

We don't know how many patients are worse off following chemotherapy and radiation treatment. As previously stated, a recent study in the United Kingdom of 23,228 patients with breast cancer and 9634 patients with non-small cell lung cancer showed 8.4% of patients undergoing treatment for lung cancer, and 2.4% of those being treated for breast cancer, died within a month of commencing therapy.

17.7 Adapting to drug treatment

For nearly 200,000 years our bodies have had to neutralize toxins from other forms of life such as plants, fungi, bacteria, and insects. There are more bacteria cells on our skins and in our digestive systems than there are normal cells in our bodies. Bacteria secrete a range of chemicals daily, some of which are toxic. We have the capability to neutralize a variety of dangerous chemicals.

As part of our natural defense, we have processes that pump undesired chemicals out of cells through the outer cell membrane. Cancer cells have access to this technology, which may be used as is, or may be adapted to stave off severe and unrelenting attack from drugs. There are several ways drugs can be rendered ineffective by cells:

- They may improve the efficiency or diversity of inbuilt pumps to elute drug molecules out of the cytoplasm into the space between cells, thereby lowering their concentrations inside to sub-toxic levels.
- Drug molecules may be modified in some way so that they are no longer able to bind to their targets, thereby reducing their effectiveness.
- Drug targets may be modified so that drug molecules are no longer able to bind to them.
- Mutations to a different protein may counter the action of an effective drug.

17.8 Targeted therapy

The current system of classification by tissue type and appearance has its limitations. In theory, it is possible to use the gene mutation and expression profiles of a given cancer to identify culpable genes, and customize drug treatment accordingly. Currently there are major challenges to achieving this:

- A gene that is a driver for one tumor may not be for another.
- No single driver gene has been implicated in all forms of cancer.
- No two tumors have the same genetic defects.

- Tumors are made up of multiple clones of cells.
- Among hundreds or even thousands of mutations it is difficult to distinguish driver mutations from passenger mutations.
- A single characteristic of a cancer cell may be achieved by the actions of several genes.
- A single gene may influence more than one characteristic of a cell.
- There are a limited number of affordable target-specific drugs available.

New drugs to launch attacks against vulnerable targets peculiar to a specific form of cancer are under development. As is the case regarding the development of drugs, major factors are cost, time to market, and market size.

17.9 Antiangiogenic drugs

All growing solid tumors require the extension of an existing network of vessels to supply them with blood to sustain expansion. Inhibitors of angiogenesis form a unique group of anticancer drugs as they inhibit the growth of blood vessels, rather than the direct division of tumor cells. The restriction of blood flow to any cluster of cells, regardless of tissue type or mutation profile, stops them from growing. Thus, angiogenic inhibitor therapy may not necessarily kill tumor cells, but puts a halt to their growth. Importantly though, it puts a brake on the growth of metastatic tumors. The problem with targeting angiogenesis is that patients may have to take drugs for extended periods of time. Normal cells already have their blood supply in place and so are unlikely to be affected by inhibitors of angiogenesis. However, side effects of treatment include bleeding, clots in arteries, and hypertension.

Targets of antiangiogenic drugs include:

- Growth factors, such as VEGF, that initiate the growth of blood vessels
- Receptors of growth factors that initiate the growth of blood vessels
- Protein involved in the passing of signals that initiate the growth of blood vessels

Specially prepared monoclonal antibodies to VEGF have been used as drugs. In some cancers, inhibitors of angiogenesis have been observed to be more effective when used in conjunction with additional therapies such as chemotherapy of a particular type. It has been suggested that angiogenic drugs work by somehow improving the network of blood vessels supplying tumors, thereby facilitating better delivery of other anticancer agents.

17.10 Immunotherapy

The aim of immunotherapy is to coerce the immune system into killing tumor cells. This typically involves giving it a helping hand by ramping up the production of killer cells, or by countering measures taken by cancer cells to suppress it. Immunotherapy may be complemented by chemotherapy and radiotherapy, both of which increase the release of antigens formed by the digestion of dead tumor cells, which in turn increases the possibility and intensity of an immune response. The advantages of immunotherapy over other forms of treatment are:

- It is specifically targeted at tumor cells.
- It kills both dividing and non-dividing tumor cells.
- Unlike other forms of therapy, the immune system has the potential to adapt as tumors evolve.

Unlike chemotherapy and radiation treatment, immunotherapy is specifically targeted at antigens on tumor cells. It makes no difference whether targeted cells are dividing or not. As new clones of tumor cells evolve, the immune system has the capability to keep up by increasing the production of clones of target-specific T-cells. Therein lies the potential for long-term cures, as opposed to disease-free cures. Results of clinical trials using combinations of immunotherapies show great promise. For example, in one trial nearly 90% of patients with advanced melanoma treated with a combination of two immunotherapy drugs were alive two years later, compared to an expected survival rate of 15% following conventional therapy. There are many ways to coax the immune system into action.

17.11 Cytokine treatment

Because they act to stimulate an immune response, cytokines are often administered in combination with immunotherapies. However, they also promote chronic inflammation, which is associated with the development of several types of cancer. Drugs designed to reduce the levels of inflammatory cytokines in and around tumors are currently being developed. Cytokines are known to activate immune cells such as natural killer cells, cytotoxic T-cells, and dendritic cells, and to target tumor cells directly. For example, tumor necrosis factor-alpha and interferon alpha both interact with tumor cells, and induce them to either commit suicide or stop growing. Cytokine therapy is approved for clinical use in the treatment of cancers such as chronic myelogenous leukemia, melanoma, kidney cancer, hairy cell leukemia, follicular non-Hodgkin's lymphoma, and Kaposi's sarcoma.

17.12 Cancer vaccines

The wonderful thing about our immune system is that it remembers previous infections, so when a second invasion of a pathogen occurs, a quicker response is evoked. Not so much a once bitten twice shy policy, but more of a once savaged twice paranoid approach. This is why vaccines work, and why they work so well. In theory, vaccines can be produced from antigens originating from a patient's own tumor cells, but getting this method to work consistently is proving troublesome.

The earliest known form of vaccination, called variolation, was first practiced in China in the tenth-century. It was administered by inoculating healthy individuals with extracts from smallpox lesions. Those that survived to tell the tale were protected against later more serious exposures to the pathogen. In 1796, an English physician by the name of Edward Jenner inoculated an eight-year-old boy with an extract from cowpox pustules of cows. After recovering from this ordeal, Dr. Jenner, who was on a roll, infected the youth with smallpox. As suspected by the intrepid doctor, the boy did not become infected with the disease. What would current health and safety say about such practices? The exposure of the immune system to a mild form of a pathogen builds up resistance that helps to fight off more serious future exposures. The term vaccination is derived from vacca, the Latin word for cow. Vaccinations are the single invention that has saved more lives than any other in the history of medicine.

As long ago as the 1890s, a New York-based surgeon by the name of William Coley noticed that a patient's cancer regressed following a bacterial infection. He then infected another patient who had inoperable cancer with the same bacteria. To his surprise, not only did the tumor regress, the patient lived for another 26 years. In total, Dr. Coley and other physicians treated over 1000 cancer patients in a similar manner, with varying degrees of success. Over the ensuing years Coley's method was practiced with decreasing frequency until it became virtually forgotten. Dr. Coley has come to be regarded as the father of the rapidly developing field of cancer immunotherapy.

In 1904, a large tumor of the cervix in an Italian woman regressed following a vaccination for rabies administered after a dog bit her, a case of once bitten once regressed. She managed to live cancer-free for another eight years. Soon thereafter several other Italian patients with cervical cancer were inoculated with the same vaccine. Although tumor regression was observed in some patients, they all eventually relapsed and died.

The bacille Calmette-Guerin (BCG) vaccination, which protects against tuberculosis, and which is considered a potent activator of an immune response, was approved by the FDA as a first-line form of treatment for early bladder cancer in 1990. The first cancer vaccine for commercial use was approved in 2010, to treat metastatic prostate cancer. Other vaccines for cancers of the brain, breast, and lung are in clinical trials. Of all areas of modern medicine concerning the treatment of cancer, turning the immune system against cancer cells perhaps holds the greatest promise.

17.13 Tumor cell vaccines

Initial attempts at producing vaccines for cancer involved breaking up tumor cells and injecting the crude mixture back into patients. A more recent approach has involved the use of whole tumor cells that are weakened by exposure to radiation before being injected back into patients. The thinking behind this approach is that damaged tumor cells die in the body, and are subsequently destroyed by the immune system, causing the release of a number of different antigens. These are subsequently engulfed by antigen-presenting cells, such as dendritic cells, and put on display for the attention of B-cells and cytotoxic T-cells. The greater the number of tumor antigens released by the digestion of dead tumor cells, the greater the chances of an immune response being invoked. Such a response may be boosted by the injection of cytokines, which have been shown to dramatically improve the efficacy of vaccinations. An advantage of using whole tumor cell vaccines is that antigens do not have to be purified, significantly reducing costs.

Today with a better understanding of how it works, it is believed an immune surge following infection or vaccination, possibly in conjunction with coercion to attack, is responsible for the regression of tumors. The existence of molecules that suppress an immune response may help to explain why early attempts at stimulating the immune system to kill cancer cells using vaccines failed. Although immune responses may have started, they could have been quelled by the release of suppressive molecules by cancer cells. The addition of checkpoint inhibitors to vaccines offers a potential means of overcoming this hurdle. Clinical trials are under way to test this possibility. Checkpoint inhibitors are explained below.

17.14 Targeting breast cancer

In one form of breast cancer the HER2 gene, which codes for the HER2 receptor, is over expressed. Roughly 20,000 receptors are found in the outer membrane of normal, healthy breast cells. In approximately 25% of primary breast cancers, the HER2 gene is over expressed resulting in as many as two million receptors.

The result of this is an amplification of the growth hormone effect that promotes uncontrolled cell division. Ostensibly the immune system does not attack these abnormal cells because the HER2 receptor is part of self. We see here a limitation of the immune response in that it fails to protect against overexpressed molecules that drive cancer development, even when they are abundant and very visible on the outer cell membrane. Attacks on molecules that are self are potentially very dangerous, as they cause autoimmune diseases such as lupus and rheumatoid arthritis.

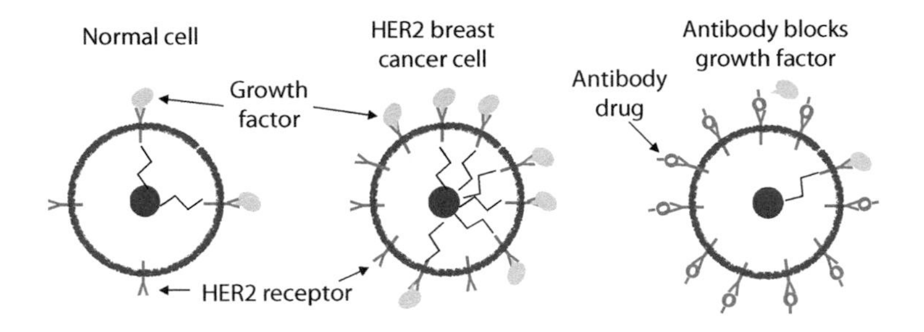

Figure 17.2 Excess HER2 receptors increase signaling of cell division. Antibody drug stops growth factor signaling by blocking access to receptors.

The drug Trastuzumab (Herceptin), which is a monoclonal antibody to the HER2 receptor, binds to it and blocks the attachment of growth hormones, thus preventing the signaling of cell division, as shown in Figure 17.2. Because of the large number of receptors on abnormal cells, Trastuzumab (Herceptin) preferentially binds to them. Since it is a foreign molecule, and therefore an antigen, the binding of Trastuzumab (Herceptin) to cancer cells attracts local antibodies and killer cells, leading to their destruction.

Scientists have been able to produce monoclonal antibodies such as Trastuzumab (Herceptin) since the 1970s. They are used to treat a variety of medical conditions including cancer, for which they typically bind to molecules, such as those involved in signaling. They thereby suppress their normal function and promote destruction by the immune system. At present the FDA has approved the use of 20 or so monoclonal antibodies for the treatment of cancer. They can be used to treat cancers such as diffuse large B-cell lymphomas, chronic lymphocytic leukemia, and relapsed or refractory B-cell non-Hodgkin's lymphomas.

17.15 Adoptive cell transfer

Progress is being made with an experimental form of immunotherapy referred to as adoptive cell transfer. It is a common observation that in many forms of cancer, killer cells of the immune system such as macrophages, neutrophils, and cytotoxic T-cells become associated with the tumor microenvironment. Although this demonstrates the ability to recognize tumor cells, it also shows a lack of capability to destroy them. It is assumed molecules released by tumor cells somehow suppress the attacking machinery of killer cells. In one method of treatment, tumor-associated T-cells are harvested and cloned in a laboratory. They are then energized by exposure to cytokines and infused back into the patient's bloodstream. The aim of this approach is to provide a boost to the immune system to help it launch an overwhelming attack. Clinical trials have demonstrated this method to be effective in completely eradicating tumors in some patients with very advanced cancer that were primarily cancers of the blood. To improve the effectiveness of adoptive cell transfer, it may be necessary to first obliterate the patient's immune system with high doses of chemotherapeutic drugs to eliminate suppressed immune cells and to create room for the new fired-up cells.

A variation on the cloning theme is production of genetically engineered T-cells in a laboratory that explicitly express antibody receptor molecules which recognize antigens specially prepared from tumor cells. In this manner, clones of T-cells can be produced that are armed with antibodies and are specific to a given tumor. The antibody receptors permit the modified T-cells to bind to antigens on cancer cells and kill them. Specially chosen viruses are used as vectors to incorporate genes into the DNA of T-cells. Genetic engineering offers the potential to expand the range of cancers that can be treated with adoptive cell therapy such as leukemia, lymphoma, neuroblastoma, and synovial cell sarcoma.

17.16 Therapeutic antibodies

Therapeutic antibodies are specially prepared by chemically attaching toxic molecules to antibodies raised against antigens from cancer cells. The intention is for such guided conjugates to selectively bind to and become incorporated into cancer cells, following which their toxic payload exerts its mortal effect. The toxin may be a radioactive compound, a drug, or a biological molecule. On binding, some antibodies can induce cancer cells to undergo apoptosis without the need for a toxic payload. Several therapeutic antibodies are currently being used to treat different forms of cancers such as breast cancer, Hodgkin's lymphoma, and a type of non-Hodgkin's B-cell lymphoma.

17.17 Checkpoint inhibitors

To effectively eliminate tumor cells, the immune system needs to be able to generate an appropriate response. Killer cells need to be mass-produced and switched on when a response is initiated, and switched off once it has run its course. There are pathways that are not well understood that regulate the state of cytotoxic T-cells so that they are not in a permanent active state. This helps to keep autoimmune disorders in check. Immune checkpoints refer to a range of pathways that can switch the activity of T-cells on and off. A balance between co-stimulatory and inhibitory signals regulates the amplitude of the response of T-cells. Cancer cells sometimes find ways to exploit checkpoints to suppress attacks from the immune system, for example, by dysregulating the expression of immune-checkpoint proteins. The use of drugs to block this suppression holds a lot of promise, as a new form of cancer treatment that keeps killer cells active, and switches on those that are inactive. Agonists of co-stimulatory receptors or antagonists of inhibitory signals, both of which result in the amplification of T-cell responses, are primary candidates for drug targets. They have been shown to be helpful in treating several types of cancer, including melanoma, Hodgkin's lymphoma, non-small cell lung cancer, kidney cancer, as well as head and neck cancers.

Our adaptive immune system comes with millions of clones of cytotoxic T-cells, each with its own unique antibody receptor that by chance happen to recognize specific antigens on cancer cells. These antibody receptors routinely work in conjunction with other regulatory receptors such as CD28 that prompts T-cells to attack, and CTLA-4 that prompts them to stand down. The use of monoclonal antibodies to CTLA-4 has been tested in clinical trials, and was approved by the FDA in 2011 for the treatment of advanced melanomas.

PD-1 is a checkpoint protein of T-cells that normally acts as a switch to regulate their activity. PD-L1 is a protein that can flick the switch to an off setting. When a killer T-cell with PD-1 binds to another cell with PD-L1, it is signaled not to attack. Some cancer cells have large amounts of PD-L1, which protect them from being wiped out. It is possible to produce clones of antibodies to PD-1 or PD-L1 that blocks their association, and in so doing permit attacks to take place. This is illustrated in Figure 17.3.

Immune checkpoint therapy, which targets regulatory pathways in T-cells, is an important recent clinical advance, but it does not always work. For some, it works for a while and then stops. We don't know why. Between 20% and 40% of patients respond well to treatment, which frequently elicits remission. When checkpoint inhibitors work, they do so well, effecting long periods of remission bordering on cures that persist long after the cessation of treatment.

A major concern with the activation of killer cells is that they may attack normal cells and cause unpredictable side effects. Some already noted are

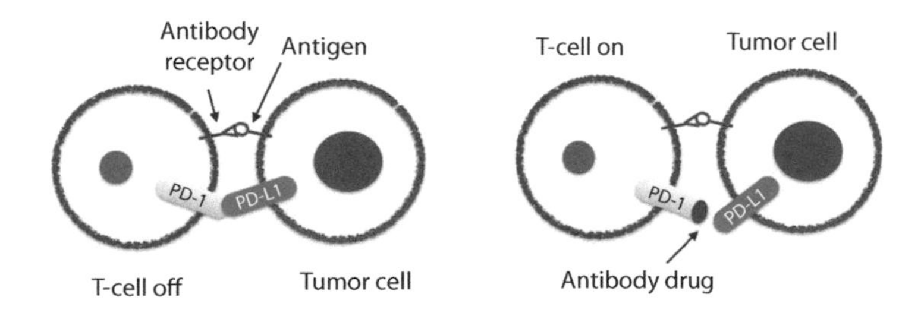

Figure 17.3 The binding of PD-L1 of tumor cell to PD-1 of T-cell switches it off, preventing attack. Antibody drug blocks switching off, enabling attack.

fatigue, nausea, skin rash, loss of appetite, and itching. Checkpoint inhibitors tend to work well on tumors with many mutations, such as melanomas and cancers of the lung and bladder, presumably because these produce more antigens for T-cells to attack. The way forward for this form of drug treatment is dependent on the development of a better understanding of checkpoint regulatory pathways and their role in creating a comfortable microenvironment for tumors.

17.18 Summary

Lifestyle choices, infections and inherited genetic predispositions are strong influences on the risk of developing cancer. It is clear we cannot expect to treat cancer the same way we treat other diseases, because each occurrence is unique and each patient is unique. We also cannot, at this point in time, expect to achieve similar results in the treatment of cancer as we achieve in the treatment of other diseases. We know enough about cancers now to appreciate why it is so difficult to cure them. We also have an appreciation of what we need to do to find cures for cancer. There isn't ever likely to be a single cure, because there isn't a single common cause. Chemotherapy and radiation therapy aren't real cures for cancer. Drug therapies targeted at corrupted signaling pathways, and therapies targeted at boosting the immune system offer the best way forward. These need to be customized and possibly combined to suit the genetic profiles of patients and tumors on a case-by-case basis. For this to occur, we need routine diagnosis of tumors based on gene mutation and gene expression profiles.

The development of immunity to treatment, and consequently the occurrence of relapses, can be addressed by using multiple drugs targeting multiple sites in pathways. Metastasis can be delayed through drugs that suppress angiogenesis. We still have a lot more to learn about the immune system to supplement the discovery of new, effective drugs and treatment. We need to develop a better understanding of biological pathways, particularly signaling pathways, and how they are corrupted for each tumor. We also need a better understanding of how:

- The innate and adaptive immune systems work
- Tumor cells evade destruction by the immune system
- Cell division is controlled
- Checkpoints of the cell cycle are bypassed

The curing of cancer is within our capability, but we need to find weak spots and aim drugs and the immune system at them.

Index